Anticipatory Ethics and The Use of CRISPR in Humans

Michael W. Nestor • Richard L. Wilson

Anticipatory Ethics and The Use of CRISPR in Humans

Springer

Michael W. Nestor
Baltimore, MD, USA

Richard L. Wilson
Philosophy and Religious Studies,
Computer and Information Sciences
Towson University
Towson, MD, USA

ISBN 978-3-030-98367-3 ISBN 978-3-030-98368-0 (eBook)
https://doi.org/10.1007/978-3-030-98368-0

This Springer imprint is published by the registered company Springer Nature Switzerland AG
The registered company address is: Gewerbestrasse 11, 6330 Cham, Switzerland

Preface

It is important to state at the outset what this book is and is not. We intend this text to be a brief and non-exhaustive introduction to the anticipatory ethical analysis of the use of CRISPR in humans. This topic is vast with many implications, so we have focused our discussion on a narrow set of topics within CRISPR-related research to spark additional conversation and analysis. We do not intend this book to be an exhaustive review of either CRISPR technology or anticipatory ethics but an applied exercise in reviewing some topics that may present an ethical quandary for future study.

It is essential to address the conceptual framework superimposed on anticipatory ethics by Carter et al. (2009) in their article entitled, "Scare-mongering and the anticipatory ethics of experimental technologies." In this article, Carter et al. state:

> Scare mongering is a persistent hazard of anticipatory ethics (Hall and Carter, 2007). In striving to anticipate all potentially adverse consequences of emerging technologies, bioethicists run the risk of highlighting implausible and unlikely adverse effects. This is often, in part, and understandable response to proponents' exaggerations of the likely benefits of the technologies. (Carter et al., 2009)

We accept this as a fair and balanced assessment of some aspects of anticipatory ethics. For this reason, our book is intended to be a framework for developing an ongoing discussion, case study, and analysis in a field that we recognize is rapidly changing. In the course of the conception and composition of this book (originally conceived in 2016), two Nobel Prizes have been awarded for the development of CRISPR. Additionally, we have observed the field advance exponentially in the interim. We have watched CRISPR move from a technique used in only the most advanced and well-equipped laboratories to one almost as ubiquitous and cost-effective as polymerase chain reaction.

That said, the idea that anticipatory ethics is biased toward scaremongering by highlighting potentially "implausible and unlikely adverse effects" is at its core a *ceteris paribus* logical fallacy that relies on the idea that science is not a dynamic and ever-changing empirical endeavor. This is especially true for the biological sciences, where there are few established laws of nature and where new technologies can layer onto old technologies to reveal effects via a retrospective analysis that may not be readily observable in the present. As a retrospective, analysis of scientific data or the scientific literature can change the interpretation of data and hypotheses, so to can *prospective* analysis and forecasting based on currently available data—as is the case in developing an anticipatory ethical analysis.

We intend to be careful in this book not to guide the reader towards our interpretation of the potential outcomes for CRISPR in humans by developing a case around scaremongering. One of us is a biomedical scientist who has used CRISPR in his own research, and we both agree that it is an important research tool. The overarching question, which wraps around many essential laboratory research tools, is whether CRISPR is appropriate for use outside of a laboratory setting and in humans in a potentially uncontrolled environment?

If we agree, as we do, that it may be appropriate to use CRISPR in *some* cases in humans outside of a controlled laboratory environment, then what are the appropriate use case and policy frameworks that need to be in place to bolster the use of this technology?

The development of this book before and during the COVID-19 pandemic has underscored the need for international cooperation and coordination from a policy perspective around the use and dissemination of biomedical technology. Before the COVID-19 pandemic, although we would see *some* agreement with us on this point, we believe that after experiencing this ongoing pandemic, *almost all* reading this will intrinsically understand the need for the development of policy and governance that is inherently anticipatory. We accept the risk that some may see this manuscript as fearmongering or potentially developing a straw man argument for a greater reward. That is, the majority will use this as an opportunity to begin to think more profoundly and long-term about the implications of this new generation of advanced techniques when used in biomedical research.

We also need to acknowledge the difficulty of writing a book of this nature. The nature of gene editing in general, and CRISPR in particular, is an extraordinarily fast-paced environment where the landscape changes seemingly daily. It is highly likely that some portion of it may be out of date by the time you read this book. Writing a book like this required us to change our perspective daily as new research manuscripts were published or policy positions were announced. Hopefully, the reader understands that this was much like hitting a moving target while moving ourselves. Our goal in this book was to develop a discussion rooted in an anticipatory ethical framework that reflects a theoretical dialog between scientists and philosophers. We wanted to capture what that conversation might look like and what issues or ideas would arise from a de-siloed, interdisciplinary exchange.

One might ask, why the emphasis on the lived experience of future persons concerning technology? Didn't inventing the electric lightbulb alter the lived experiences of future persons? Should we not have pursued research into the lightbulb for fear that it might change how future persons live? Isn't that the case with every new technology, in that every new technology alters the lived experience of future persons? These are valid questions, and it is essential to underscore why the use of CRISPR, specifically in conjunction with germline gene editing, and even more specifically to alter genes that play a role in cognition and the brain, is a special case. We will discuss this in more depth in Chap. 4. Somatic CRISPR editing, like the invention of the electric lightbulb, may alter some aspects of the lived experience of future persons. However, in our opinion, somatic CRISPR editing will not change the *very essence* of how we define ourselves as humans and the concept of personhood itself.

On the other hand, germline CRISPR editing of genes involved in human cognition can completely alter what it means to engage in the human experience in a way that correcting a set of genes that can reverse sickle cell anemia does not. In this way, any technology with such a disproportionate impact should be subject to discussion and debate, and policies should be built to define the guardrails for that technology. We recognize that scientifically it is not currently possible to do such work, but that does not mean it is improbable in the future. This is the unique opportunity anticipatory ethics provides. It allows us to imaginatively construct a future world in the present and ask ourselves, from an ethical perspective, is this a world in which we would like to live? By using the past and present to project into the future, we can develop policies and processes. This framework can be used to bring the public, scientists, and the government closer together and develop an understanding across these stakeholders about what we want our future to look like. By having a dialogue now and together while including the most comprehensive set of stakeholders possible, we can increase engagement with each other while developing trust. The strength of that trust between stakeholders from the public, scientific, and governmental spheres will strengthen the bonds between those institutions. This will keep us on a steady course for the continued maturation and innovation of our worldwide scientific research and development enterprise. This is the same enterprise that itself has given us the promise of CRISPR.

Baltimore, MD, USA Michael W. Nestor
Towson, MD, USA Richard L. Wilson

Reference

Carter, A., Bartlett, P., & Hall, W. (2009). Scare-mongering and the anticipatory ethics of experimental technologies. *The American Journal of Bioethics, 9*(5), 47–48. https://doi.org/10.1080/15265160902788736

Acknowledgments

Thank you to Elizabeth Samit for additional assistance with Chap. 1. Thank you to Dr. Michael S. Kuefner for additional assistance with Chaps. 7 and 8.

Contents

About the Authors

Michael W. Nestor received his PhD in neuroscience from the University of Maryland, School of Medicine, and has dual certificates in regulatory affairs—medical devices and pharmaceuticals from RAPS and a certificate in project management from the University of California, Irvine. He completed postdoctoral fellowships at the National Institutes of Health and The New York Stem Cell Foundation, where he was also a staff scientist. Concurrently, Michael was an NIH IRACDA Fellow at Rutgers University, focused on teaching in minority-serving institutions and helping increase the participation of traditionally underrepresented groups in science. Dr. Nestor was awarded an American Association for the Advancement of Science, Science and Technology Policy Fellowship and served in the executive branch.

A neuroscientist with 20 years of research experience, Dr. Nestor was director of Neural Stem Cell Research at the Hussman Institute for Autism, where he led his laboratory studying autism by creating brain organoids from human induced pluripotent stem cells. His lab developed multiplexed high-throughput CRISPR and drug-screening platforms using human stem cell derived brain organoids. Dr. Nestor ran his own human stem cell consulting company, Synapstem, and acted as a venture advisor to the UM Momentum Fund and the Abell Foundation and as co-chair of the Neural Stem Cell Working Group at the University of Maryland, School of Medicine.

Richard L. Wilson teaches philosophy and ethics in the Philosophy and Computer and Information Sciences Departments at Towson University. He is a senior research scholar in the Hoffberger Center for Professional Ethics at the University of Baltimore. Professor Wilson specializes in applied ethics and has taught a wide variety of applied ethics classes. As a specialist in applied ethics, he has taught courses and published a variety of books and articles in engineering ethics, computer science ethics, biomedical ethics, business ethics, media ethics, and environmental ethics. Recent publications include a book, *Glossary of Cyber Warfare, Cyber Crime and Cyber Security,* with a variety of papers published including *3D Printing and Anticipatory Business Ethics*, *An Anticipatory Ethical Analysis of*

Offensive Cyberspace Operations, *The Use of Facial Recognition in China's Social Credit System: An Anticipatory Ethical Analysis*, *Anticipatory Medical Ethics: Pharmaceuticals and Nanotechnology*, and *The Ethics of Autonomy and Lethality*. His most recent publications include*: Anticipatory Ethics as a Method for Teaching Engineering Ethics* in the ASEE Proceedings and *Smart Cities and Cyber Security: Ethical and Anticipated Ethical Concerns*.

List of Figures

Chapter 1
Brief Introduction to CRISPR Cas9

Editing of the human genome using clustered regularly interspaced short palindromic repeats (CRISPR) with CRISPR-associated protein 9 (Cas9) (typically referred to as simply CRISPR) has already changed the nature of how research scientists, clinicians, and ordinary people think about human disease and health. The nature of CRISPR/Cas9 technology is changing as rapidly as it can be developed and put into clinical trials and, as a gene-editing technique, it represents a revolution in science and medicine. It is no wonder that recent Nobel prizes have been awarded to Jennifer Doudna and Emmanuelle Charpentier, as the technique they developed for efficient gene editing is groundbreaking for areas of biology well beyond humans. Applied to human beings, the application of CRISPR/Cas9 allows for rethinking the meaning of genetic conditions and how to treat them to alleviate suffering and potentially ameliorate some diseases.

The purpose of this chapter is to provide you with the following: a brief historical overview of CRISPR/Cas9, description of the technology, current animal/human models for treating disease as linked to CRISPR/Cas9, existing clinical applications for CRISPR/Cas9 in humans, and present (and future) challenges as related to CRISPR/Cas9 utilization.

1.1 Historical Overview

Nearly two decades before the development of CRISPR/Cas9, medical research scientists studying acquired immune responses in bacteria and archaea (a class of single-celled organisms) discovered the existence of interspaced, short, palindromic DNA repeats. In 1987—while performing a metabolism-focused analysis of the genome of the common bacteria *E. coli*—a group of researchers discovered an unusual repetitive DNA sequence that was foundational to all subsequent scientific research, enabling the development of CRISPR (Ishino et al., 2018).

M. W. Nestor, R. L. Wilson, *Anticipatory Ethics and The Use of CRISPR in Humans*, https://doi.org/10.1007/978-3-030-98368-0_1

The Spanish microbiologist Francisco Mojico (of the *University of Alicante*, Spain) in 1993 published findings pertaining to the functional mechanism of the repetitive DNA sequences (which he called tandem repeats or *TREPS*). Although underappreciated in its relevance to biomedicine in the mid-1990s, other scientists' investigations stemming from Mojico's findings led to the understanding that these repeats played a critical role in the adaptive immune system (Lander, 2016).

Specifically, the spacers between the repeats were found to match those of formerly invading viruses that were now no longer able to invade single-celled organisms due to a protective immune response against the invader. Mojico and others coined the term CRISPR (which was first used in 2002 in an article by Rudd Jansen published in *Molecular Microbiology*) (CRISPR Timeline, 2016). In this Jansen article, the existence of *cas* genes adjacent to CRISPR loci was first described (Jansen et al., 2002). It is *cas* genes that encode for CRISPR-associated proteins that are fundamental to the functional complexes that enable the action of CRISPR/Cas9 (Makarova et al., 2019).

During this same time period, the mapping of the human genome was nearing completion (with completion occurring in 2003). Biotechnology for drug delivery (as well as diagnostics) was enabling novel treatments for chronic health disorders previously considered fatal. For this concise overview, we will skip to 2005. In that year, lab-based researchers demonstrated that introducing new genetic elements into organisms could alter the functioning of the immune system response in one-celled organisms. Notably, while other *cas* genes (proteins) performed certain functions, *cas9* acts on bacteriophages—so it is the only one required for CRISPR/Cas9 targeting to produce a viral protective immune response where none previously existed. In this context, *cas9* acts like scissors that can "clip" molecular strands of DNA (Gasiunas et al., 2012; Jinek et al., 2012). The CRISPR/Cas9 system does this clipping by deletion of two or more nucleotides. When the cell uses the intrinsic machinery that has evolved to repair the DNA (called non-homologous DNA end-joining), the process results in a frameshift and resultant silencing of the edited genes (Guo et al., 2018). This discovery (along with the subsequent recognition of CRISPR/Cas9 RNA roles) unlocked the potential of *intentional* gene-editing to generate a protective response. Colloquially, CRISPR has become an acronym representing the entire CRISPR/Cas9 system (Vidyasagar, 2018).

There is an ongoing patent dispute between Jennifer Doudna and Feng Zhang, who, together with Emmanuelle Charpentier, was the first to show broad gene-editing applications for CRISPR/Cas9. This work resulted in the awarding of a Nobel Prize to Doudna and Charpentier in 2020. The patent for this technology—enabling its commercialization for use in the biotechnology and pharmaceutical marketplace—was awarded in 2018. That awarded patent (although challenged by researchers at the University of California) was granted to the Broad Institute for its CRISPR technology to make targeted changes to the genomes of eukaryotes (which are organisms consisting of a cell [or cells] in which the genetic material is DNA in the form of chromosomes contained within a distinct nucleus) (Ledford, 2018).

The details of this story are fascinating and important for the broader scientific enterprise but ultimately beyond the scope of this manuscript. Importantly, in 2013,

Zhang's group used CRISPR/Cas9 to edit mouse and human cells at the Broad Institute at the Massachusetts Institute of Technology (MIT) (Cohen, 2017). CRISPR/Cas9 was now poised to become a cutting-edge (and revolutionary) medical tool.

Currently, three different medical uses for CRISPR/Cas9 are generally recognized (besides various non-medical uses, such as agriculturally related and veterinary uses). The first medical use is for research purposes (such as discovering the various roles played by each gene or how an altered gene can impact physiological processes). The second is for developing treatments for diseases (such as cancers). The third is for the prevention of conditions (including genetic conditions).

CRISPR/Cas9—albeit considered the most accurate and predictable current gene-editing technology—is not the only gene-editing technology in existence. A review of the other major current gene-editing technologies (such as transcription activator-like effector nucleases [TALEN] or zinc finger nucleases [ZFN]) is beyond the scope of this manuscript. It is essential to underscore that there is a multitude of gene-editing technologies, some of which may have advantages over CRISPR/Cas9 in specific scientific applications and contexts (Jain et al., 2021). Historically, the most critical tools for genome editing have been enzymes like ZFNs, TALENs, and engineered mega-nucleases, which are being eclipsed by the exponential growth surrounding the use of CRISPR/Cas9 (Gao et al., 2020). CRISPR has the broadest application potential, and for use in humans, has a significant advantage due to the specificity and fidelity of its ability to cleave DNA. This makes it most applicable to medical biotechnology approaches focused on ameliorating human disease. The unique ability. of CRISPR/Cas9 as a delivery tool is that it is based on the elements of the bacterial and archaeal mechanism responsible for acquired immunity against bacteriophage infections and transfer of foreign (e.g., viral) genetic material *and* is more precise (Czarnek & Bereta, 2016).

The CRISPR/Cas9 genome-editing tool is adopted from the adaptive immune defense system of *prokaryotes*. (in which bacteria or archaea use the CRISPR/Cas9 system to cut and destroy the invaded DNA from outside of the cells, such as the DNA of viruses). For primarily this reason, its first "real-world" utilization was in the development of knockout (and, then, knockin) mice, enabling more medically related research studies at the Phase 1 clinical trial level (Zhang, 2020).

To understand how CRISPR/Cas9 works, it is crucial to recognize that—in organisms—CRISPR/Cas systems include two classes (Class 1 and Class 2) that are based on the encoded effector proteins (such as *cas 9*). Each of these classes consists of three types (Types 1, 2, and 3); these are identifiable due to their signature enzymatic proteins (such as *cas9* for Type 2). As reported in an article in the *Journal of Bacteriology* in 2018:

> The simple architecture of the effector complexes has made class 2 CRISPR-Cas systems an attractive choice for developing a new generation of genome editing technologies. (Ishino et al., 2018)

These enzymatic proteins are responsible for making a cut in the targeted DNA sequence that is highly specific. Although there is a large family of Cas proteins,

Cas9 has been the focus for applications in humans. The CRISPR/Cas9 system/construct is delivered to a cell of interest via viral delivery using adenovirus, adeno-associated virus, or lentivirus. Other non-viral methods are also less frequently used, such as extracellular vesicles, microinjection, electroporation, or cell-penetrating peptides. Each of these delivery methods has its pros and cons, such as delivery efficiency, technical requirements, and safety concerns (Yip, 2020). The current mode of delivery that is targeted for human use in most cases is adeno-associated virus. For example, a pro for using a lentivirus to deliver the CRISPR/Cas9 to cells is high efficiency. A con is that there can be random integration into the host genome, potentially activating oncogenes and inducing tumorigenesis (Chen et al., 2017). Indeed, tumorigenesis is a general concern with CRISPR/Cas9 use in that CRISPR/Cas9 has been demonstrated to induce chromosomal truncations and deletions of tumor suppressors in a p-53-dependent process (Cullot et al., 2019). p-53 encodes for proteins that regulate the cell cycle and suppress tumor formation by regulating apoptosis and cell senescence. Cas9 was shown to inactivate p-53 in human cell lines by introducing mutations that inactivated the p-53 pathway (Enache et al., 2020).

New technologies that attempt to address some of the problems with viral delivery include chemical methods of delivery such as lipid-based nanoparticles and cell-penetrating peptides (Yip, 2020). These methods are efficient but difficult to precisely target tissues or cells of interest. CRISPR/Cas9 delivery is an important area of ongoing research and presents a challenging hurdle for further development of the technology. A more detailed review of this aspect of CRISPR/Cas9 is beyond the scope of this manuscript (Fig. 1.1).

Inserted to a viral delivery system., the CRISPR/Cas9 construct (herein referred to as simply CRISPR) is composed of:

> A target-specific sgRNA [single guide RNA], formed by the fusion of a CRISPR RNA (crRNA) and a transactivating CRISPR RNA, [which] directs the Cas9 protein to a target site for cleavage, creating a double-strand break (DSB). Because target recognition is based on RNA–DNA interactions, CRISPR/Cas9 has the advantages of the easy design of genomic targets and multiplexing over that of zinc finger nucleases (ZFNs) and transcription activator-like effector nucleases (TALENs) . In contrast with ZFNs and TALENs, which require laborious protein engineering steps for each new editing target, the Cas9 nuclease simply requires a target-specific sgRNA for each editing target. (Yip, 2020).

The major component of this system is the Cas9 protein. As stated earlier, a large family of Cas proteins (13 of them) drives the CRISPR immune response, which creates immune memory in bacteria to protect bacteria against subsequent infection (Shmakov et al., 2018). The size and specificity of these proteins are crucial to gene editing because they play a role in both the efficiency of the editing step and the size of the construct that can be delivered to the cell. Smaller Cas proteins like the recently discovered Cas-phi allow for additional proteins and other DNA repair elements to be delivered, allowing for more specific targeting and less-off target effects (Pausch et al., 2020). Cas9 is currently the standard Cas protein used and

> The target specificity of Cas9 is determined by the spacer sequence of crRNAs (~20 nucleotides) and adjacent protospacer adjacent motifs (PAMs). More precisely, the seed sequence

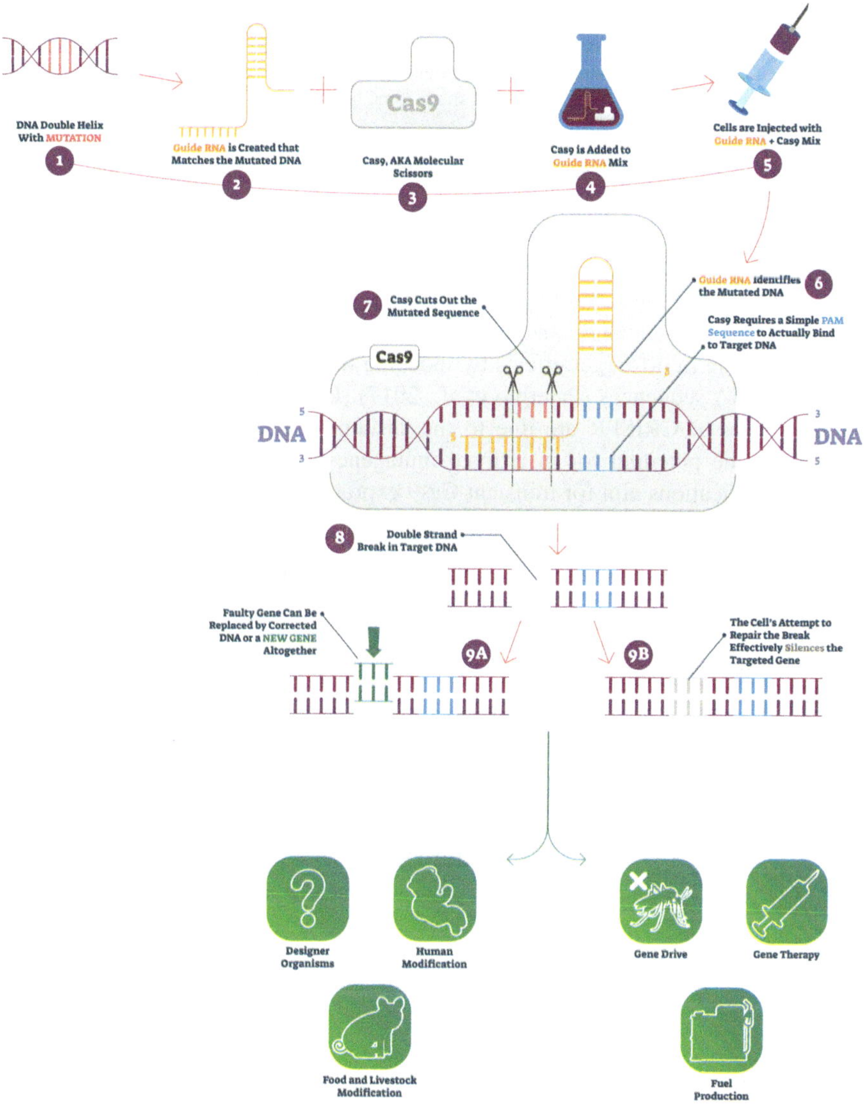

Fig. 1.1 Clustered Regularly Interspaced Short Palindromic Repeats (CRISPR) is a gene-editing technology that allows for the insertion sequences to correct DNA or silencing of targeted genes. A typical example is: (1) atypical DNA sequence with a mutation is identified and (2) gRNA is developed to match the DNA mutation, (3, 4) Cas9 is added into the gRNA mix, and (5) gRNA-Cas9 can be introduced to an in vitro system. Once in the in vitro system, (6, 7) the gRNA identifies the native gene mutation, and Cas9 enzymatically cuts the sequence, creating a (8) DNA with a double-strand break. (9a, 9b) The intrinsic machinery in the cell will repair the double-strand break, which can be used to either silence the targeted gene or introduce a new gene. This has a wide number of applications, from fuel production to plant modification to human modification

> located in the 3′ end of the spacer sequence (10–12 base pairs adjacent to the PAM) is critical for correct targeting. Cas9 will cleave only when sufficient homology is present between the seed region and the target DNA. However, off-target cleavage occurs when DNA sequences contain a few mismatches but share some homology with the seed region of the sgRNA. (Yip, 2020).

Because the PAM is the specific short sequence required for Cas9 to cut the DNA, the Cas nuclease will scan for the PAM before unraveling the DNA to create the edit. This is important because the location and presence of PAM sequences determine what genomic locations can be targeted for editing by CRISPR. The absence of a specific PAM sequence for Cas9 at the target genomic locus means that Cas9 cannot cleave that sequence. In addition, the PAM sequence. is essential because it can play a role in off-target effects by inducing the cleavage of non-fully complementary DNA sequences (Palermo et al., 2017) (Fig. 1.2). Many of the other off-target effects of CRISPR are due to the sustained expression of Cas9 in cells, increasing the potential for insertional mutagenesis (Rivera-Torres et al., 2017). Current applications aim for transient Cas9 expression and engineering of Cas9 to

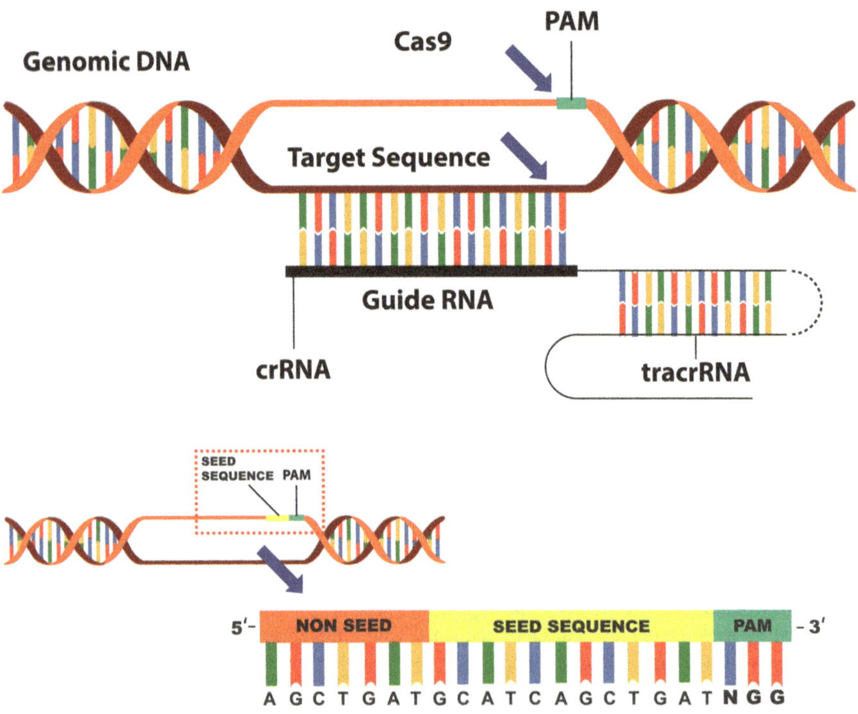

Fig. 1.2 The protospacer adjacent motif (PAM) sequence is a gene sequence that follows the sequence of DNA targeted for cleavage by Cas. It is required for Cas recognition of complementary DNA to the crRNA (short CRISPR RNA). Without the PAM, Cas cannot enzymatically cut the appropriate DNA. Cas9 sustained expression can induce unwanted off-target effects depending on the composition of the seed sequence and the PAM and can induce mutagenesis

reduce the requirement of specific PAM sequences in the genome (which we will discuss later in this chapter) at the cost of efficiency of the edits. Cas protein discovery and engineering is still an area of active research and is undergoing exponential growth as improvements on CRISPR are made.

One emerging focus of medically oriented CRISPR utilization is removing, replacing, or correcting undesirable genes that cause genetic conditions. CRISPR utilizes small RNA molecules (crRNA and tracrRNA) to precisely direct the effector nuclease—Cas9—to a specific site. Underpinning and driving CRISPR direction of Cas9 to a specific site (i.e., direction to the targeted sequence complementary to crRNA) are two functional domains of Cas9: one is HNH (a protein), and the other is RuvC (a nuclease). Each of these active catalytic domains cuts one strand of the targeted DNA sequences and forms the DNA double-strand breaks (DSBs) (Khan et al., 2018a, b).

DSBs can then be the active site for the insertion of new sequences or "corrected" sequences of DNA. Because evolution has resulted in cells developing intrinsic DNA repair mechanisms (the most common being nonhomologous DNA end joining [NHEJ] and homology-directed recombination [HDR]), these can be used to develop gene knockin/replacement or gene knockout strategies (Zhang, 2020). Additional strategies for using CRISPR in gene therapy include base editing and prime editing. Base editing relies on a modified Cas endonuclease that modifies either or both of the HNH or RuvC domains to create a Cas9 nickase (nCas9). The development of the nickase is essential because it allows for only a single-strand break (SSB) instead of the typical DSB. This allows for the creation of a fusion protein where nCas9 can be fused with cytidine deaminase or adenine deaminase enzymes such that single DNA base editing can occur (Komor et al., 2016; Gaudelli et al., 2018). These are referred to as cytosine base editors (CBEs) and adenine base editors (ABEs). They can be combined to create dual-deaminase CRISPR base editors (Grünewald et al., 2020; Sakata et al., 2020) (Fig. 1.3). CRISPR base editing has immense potential for application to human diseases caused by a single-nucleotide polymorphism (SNP). This represents a large portion of human genetic conditions, 47% of which are caused by nucleotide pair changes from A-T (adenine-thymine) to G-C (guanine-cytosine) (Rees & Liu, 2018). This approach has already been used in human cells to repair human intestinal organoids harboring mutations that result in cystic fibrosis (Geurts et al., 2020) and a mouse model of Duchenne muscular dystrophy (Ryu et al., 2018). However, one drawback to current base editing techniques is the potential for off-target effects. Prime editing has been developed to address some of these issues.

Prime editing takes advantage of the guide RNA by substituting it for a prime editing guide RNA (pegRNA). This pegRNA is linked to a primer binding site encoded by a gene-specific RNA sequence and is complementary to the targeted DNA region (Anzalone et al., 2019). This, combined with the nCas9, causes an SSB. The cleaved DNA sequence becomes a primer, binding with the targeted primer binding site in the pegRNA to synthesize a new DNA fragment *without a donor DNA template*. This allows for the substitution of any nucleotides with minimal off-target effects and has been applied to correct mutations that underlie

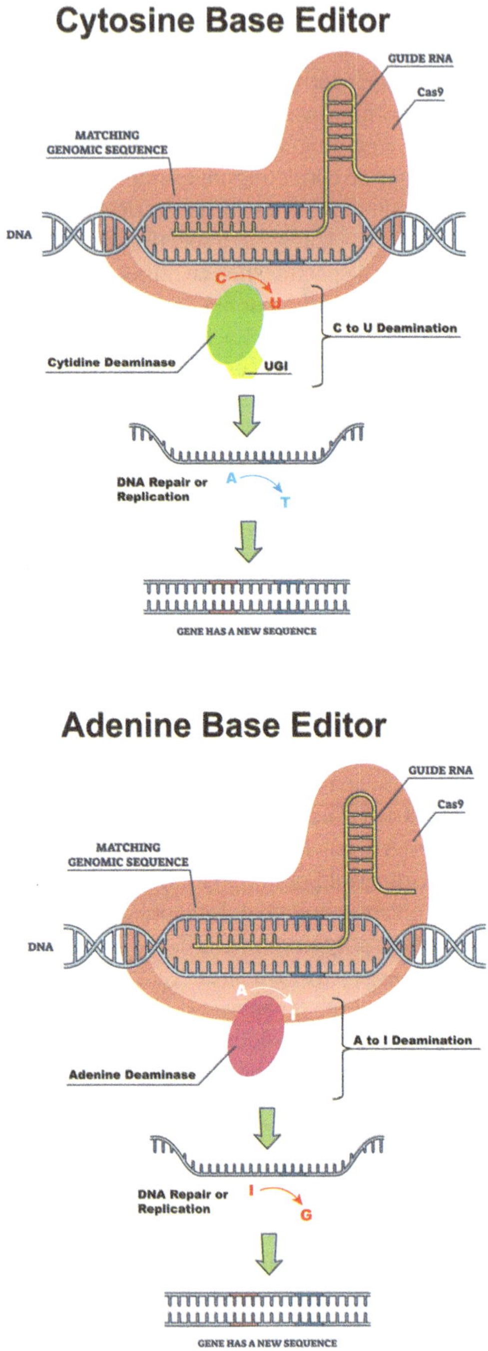

Fig. 1.3 Base editing represents a new technique for gene editing which can confer specific targeting of genes with potentially fewer off-target effects. By using modified Cas9 fused with either cytosine deaminase or adenine deaminase, the larger Cas9 complex can convert cytosine-guanine base pairs to thymine-adenine base pairs or adenine-thymine to guanine-cytosine base pairs at specific targets in the genome, allowing for correction of single-point mutations

Tay-Sachs and sickle cell diseases (Anzalone et al., 2019). Currently, the following are understood seven basic applications for CRISPR for utilization in the field of medicine:

1. Using CRISPR libraries for genetic screening (such as to predict future vulnerability to developing a specific disorder [e.g., HER2 linked to an aggressive form of breast cancer]).
2. Using CRISPR for genome editing (such as to remove a defective gene linked to a condition).
3. Transcriptional activation/suppression (within a given genome).
4. CRISPR-mediated chromatin immunoprecipitation (for clear characterization of RNAs).
5. Epigenetic editing utilizing CRISPR.
6. Live imaging of DNA and/or RNA.
7. Therapeutic options such as treatment for specific conditions.

As aptly stated by Sansbury et al. (2019) in *Communications Biology*:

> The evolution of clustered regularly interspersed palindromic repeats (CRISPR) and CRISPR-associated Cas nucleases has given researchers a novel tool with which to disable malfunctioning genes or to correct single-base mutations or small mutagenic deletions that are responsible for devastating disorders.

On the other hand—although CRISPR has already been relatively well studied in diverse human clinical treatments as of 2021—three significant challenges for its overall adoption in the medical arena are:

- The potential for toxicity and immune response of treated cells to any particular CRISPR/Cas component.
- A highly efficient, specifically targeted, and high throughput delivery method.
- Potential off-target impact (with unknown consequences) of the gene-modification.

1.2 Current Animal/Human Models to Treat Disease

The use of CRISPR is changing the conception of what is considered a pharmaceutical treatment or a medication. Not only is CRISPR a cutting-edge tool in the treatment of various genetic conditions but it is also increasing the ability to diagnose the genetic propensity for developing a future health condition. CRISPR also enables the production of mRNA vaccines (such as the mRNA vaccines against the coronavirus (publicly termed *Covid-19*).

Pharmaceutical research and development (R&D) approaches have resulted in an increased knowledge base of the impact of currently available drugs to treat various conditions. However, these are chemically composed treatments for people, independent of the person's *unique* cellular composition. In contrast, personalized medicine approaches aim to tailor treatments to match individuals' genetic and cellular backgrounds. CRISPR represents an approach to treating disease from a personalized perspective that considers each individual's genetic backbone. This same advantage can be applied to animal models of human diseases, where CRISPR can be used to edit the embryo of the animal, followed by implantation and breeding. This allows for germline editing in animal models of human conditions to understand the underlying biology and develop personalized therapeutics (Rees & Liu, 2018).

Repeated in vivo experiments performed by medical science researchers in animals (e.g., rodents) that follow successful in vitro lab-based outcomes occur before any experimental treatment (clinical) trial is undertaken in humans. The use of "knockout" mice (in which a specific gene has been eliminated or modified) as well as "knockin" mice (in which an absent genetic component is inserted) have enabled experiments to determine such findings as:

- Whether a given gene causes or fosters a health-damaging process to occur.
- Whether the absence of a genetically mediated process (such as an enzymatic process) affects physiological processes/health quality.
- Whether a specific pharmaceutical agent's effectiveness as a treatment for a health condition is genetically mediated.
- Whether a pharmaceutical agent's effect differs between individuals (e.g., mice with specific genes "knockedin" or "knockedout"), depending upon certain genetic factors.
- Whether a disorder can be prevented depending upon the existence or absence of certain genetic factors.

There is a multitude of animal models being developed using CRISPR and at a rapid pace. To survey and list the landscape of these models is beyond the scope of this manuscript. CRISPR-developed animal models are diverse. As an example, they run from mouse models of human non-small cell lung cancers (Maddalo et al., 2014) to Duchenne muscular dystrophy (Min et al., 2019), to Lateral meningocele syndrome (Canalis et al., 2018) to corneal dystrophy (Kitamoto et al., 2020). In pig, they cover conditions such as Huntington's disease (Yan et al., 2018); in African clawed frog-retinitis pigmentosa (Feehan et al., 2017). This non-exhaustive list shows the diversity of applications and animal models that can be developed for CRISPR. This does not include the in vitro models based on human stem cells or human stem cell organoids, of which there are many. Examples here include human gut organoid models of cystic fibrosis (Schwank et al., 2013), human intestinal organoid models of colorectal cancer (Matano et al., 2015), and beta-cell islet-like organoids for diabetes (Yoshihara et al., 2020). Indeed, human stem cell-derived organoid models represent the most significant current opportunity for the development of relevant CRISPR models of human disease (Kim et al., 2020; Durens et al., 2020; Lunden et al., 2020).

1.3 Current Clinical Applications for CRISPR in Humans

There is tremendous potential for applying CRISPR in the development of the class of pharmaceuticals frequently termed *biologics* and overall gene therapy as a condition treatment modality. Inherited genetic conditions were the first human health conditions targeted for the development of CRISPR clinical applications, such as the following:

- Familial conditions caused by dominant and Recessive traits
- Conditions causing specific abnormal enzymatic/hormonal processes
- Digestive and nutritionally related abnormalities
- Autoimmune disorders

1.4 Familial Conditions Caused by Dominant and Recessive Traits

1.4.1 Blood Conditions

An abnormal gene at conception (passed on the X or Y chromosome) may result (depending upon whether a dominant or recessive trait) in the potential to affect the embryo. Hemophilia is a well-known example of one such disorder, and hemophiliacs' lack of blood-clotting ability can be life-threatening. Although the current standard treatment for Hemophilia A or B is self-infusion of the missing clotting factor, it was only in 1992 that the FDA approved the use of a genetically engineered clotting factor. This enabled using DNA to replace the previously sole permission by the FDA for a clotting factor delivery mechanism of human plasma. Problematically, a delivery mechanism of human plasma requires its availability from donors and the risk of transmitting an unrecognized infection to the Hemophiliac (as occurred in the US during the early years of the HIV/AIDS pandemic).

In 2016, study findings showed success in thwarting the genetic defect in Hemophilia preventing normal blood-clotting, utilizing CRISPR. The peer-reviewed article describing these positive findings reported that CRISPR gene therapy using recombinant adeno-associated virus (rAAV) has been tested in clinical trials to treat Hemophilia B with excellent results at 5-year posttreatment follow-up (Nguyen & Anegon, 2016). Significantly, the advantage of CRISPR gene therapy is that it does not need to occur more than once in an individual in many circumstances, thus eliminating the necessity for a lifetime of ongoing treatment or medication use.

Two other inherited blood disorders that are considered good candidates for gene therapy are Sickle Cell Anemia (in which red blood cells are misshapen) and Beta Thalassemia (causing dangerously low production of hemoglobin). These diseases are highly linked to debilitating complications and can be fatal (Henderson, 2021). Yet other inherited blood disorders considered candidates for gene therapy include

Fanconi Anemia (causing decreased production of all blood cell types) (Richardson et al., 2018) and Familial Erythrocytosis (causing blood clots due to excessive production of red blood cells) (Zmajkovic et al., 2018).

1.4.2 Conditions Causing Specific Abnormal Enzymatic/ Hormonal Processes

Transthyretin Amyloidosis (TA) is an inherited disorder characterized by a progressive build-up of abnormal deposits of a protein (amyloid) in the body's organs and tissues, leading to diverse untoward symptoms. A CRISPR drug (biologic) in 2021 was injected into the blood of clinical trial participants with TA; that infusion almost totally disrupted the abnormal enzymatic process causing TA (Kaiser, 2021).

The significance of this accomplishment is that it heralded a milestone in the effort by pharmaceutical firms *in tandem with* biotechnology firms to develop treatments based on messenger RNA (mRNA), which are the protein-building instructions naturally made by cells. This treatment—involving an mRNA encoding one of CRISPR's two components—represents the convergence of the biotechnology fields of CRISPR and mRNA applications (Kaiser, 2021).

1.4.3 Digestive and Nutritionally Related Abnormalities

In pernicious anemia—which is often familial (but not always)—the afflicted person's body cannot utilize vitamin B12 properly. However, vitamin B12 is essential for the proper development of red blood cells. A hereditary link is likewise often found in people with lactose intolerance (an intolerance to the enzyme commonly found in dairy products, typically resulting in diarrhea/GI cramping upon consumption of any dairy product). Meanwhile, people with inherited hypocalcemia due to a genetic defect can experience widespread physiological impacts, especially in the musculoskeletal system.

The overall adverse impact of gene- or cellular-level abnormalities resulting in the inability to tolerate, absorb, and/or utilize specific nutrients is widely recognized as contributing to future health problems. Cell engineering via CRISPR has uncovered genes that may enable a way to correct the underlying genetic abnormality to eliminate the need for any other type of lifetime treatment/management (Ha et al., 2020).

1.4.4 Autoimmune Disorders

Lupus, Crohn's Disease, and Type 1-Diabetes are just a few of the many autoimmune disorders that tend to run in families (Bhamidipati et al., 2021). Because a wide range of autoimmune diseases can negatively impact so many physiological processes and overall health, the ability to halt the overactive immune response (leading to immune system attack on the afflicted person's organs and tissues) is an active area of research.

CRISPR may enable the capacity to alter the immune response in people afflicted with certain autoimmune disorders, such that the progression to long-term adverse physiological effects is prevented. Indeed, autoimmune disorders are a significant focus of some current, NIH-funded CRISPR research studies at major US medical centers.

The five foremost treatment areas that are the focus of all CRISPR-related human clinical trials Crohn's as of 2021 are:

1. Blood conditions
2. Cancers
3. Ocular (eye) conditions
4. Chronic infections
5. Protein-folding conditions

Notably, all current CRISPR clinical trials are intended to edit specific cells or tissues without affecting sperm/eggs, so that no DNA changes would be passed onto future generations (Henderson, 2021). A current search of CRISPR-based gene therapy clinical trials worldwide shows more than 30 trials occurring worldwide covering a diverse set of genetic conditions (Zhang, 2020). A non-exhaustive review of some highlighted current clinical applications and trials for CRISPR is essential to deepen an understanding of how ubiquitous this gene-editing approach is poised to become.

1.4.5 Blood Conditions

As already described, Sickle Cell Anemia and Beta Thalassemia are CRISPR-targeted treatment areas. The approach taken to treat these two blood disorders with CRISPR technology does not directly correct the gene variants that cause disease; instead, it employs a "workaround" by increasing the body's production levels of fetal hemoglobin into the blood.

1.4.6 Cancers

The current focus of clinical trials utilizing CRISPR therapeutics to treat cancers is primarily aimed at blood cancers (such as leukemia and lymphoma) (Henderson, 2021). However, a current non-US clinical trial is aimed at lung cancer (Jakobsen, 2020).

Immunotherapy may—in the not-too-distant future—reduce (or eliminate) the need for chemotherapy. Since chemotherapy cannot distinguish cancer cells from normal cells, cancer patients treated with chemotherapy typically experience a wide array of debilitating side effects and complications (including a depressed immune system). Moreover, for some patients (such as those with end-stage cancer), the chemotherapy complications outweigh any potential health benefit.

A cutting-edge therapeutic approach termed CAR-T Cell Therapy involves altering the genes inside specific immune system cells (T-cells) to help them attack the cancer. T-cells are extracted, altered in a lab (in vitro) environment, and then re-inserted into the patient so that the T-cells can find and destroy cancer cells. CRISPR is likely to expand the potential for CAR-T Therapy due to increased precision in engineering the modified T-cells (Liu et al., 2019)

1.4.7 Ocular (Eye) Conditions

Inherited and other ocular disorders causing retinal damage leading to childhood blindness are considered CRISPR therapeutic targets. Successful early results of a CRISPR clinical trial aimed at preventing blindness due to an inherited eye disorder were reported in 2020 (Ledford, 2020). Different advantages (and disadvantages) were revealed using viral vectors compared to non-viral vectors in this application. Likewise, different routes of administration for gene therapy are also currently being investigated (Bulcha et al., 2021).

1.4.8 Chronic Infections

Infectious disease is a major focus of myriad biotechnology researchers/companies. This rapidly developing area of research has been accelerated by the race to develop vaccines against coronaviruses utilizing CRISPR technology. CRISPR gene-editing technology has the potential to transform both the diagnosis and treatment of infectious diseases (Binnie et al., 2021).

> CRISPR systems can also be targeted against pathogenic DNA or RNA sequences. This potential is being combined with innovative delivery systems to develop new therapeutic approaches to infectious diseases. (Binnie et al., 2021)

In terms of viral diagnostics, two analogous diagnostic systems have been designed to detect specific genetic sequences. One is called DETECTR (which detects DNA) (Chen et al., 2018), and the other is called SHERLOCK (which detects RNA) (Gootenberg et al., 2017). SHERLOCK and DETECTR are so sensitive that they are capable of single copy viral detection. This will be important for rapidly detecting and developing therapeutics for infectious disease at a low cost by taking advantage of CRISPR to engineer biologics that can be delivered via a vaccine format (Zamani et al., 2021).

HIV/AIDS—a zoonotic viral disease (transmitted via contact with blood or other bodily fluids such as semen)—emerged as a global pandemic in the 1980s and remains a global public health threat. Problematically, it has eluded the ability of medical researchers to develop a vaccine against it, although medication treatments to control it have continued to improve over the past 40 (or so) years. These medications primarily work by boosting the immune system and decreasing viral load to lessen transmission likelihood. When HIV-infected people adhere to their prescribed antiretroviral medication regimen *for life*, they can remain healthy. However, not everyone infected with HIV has access to—or remains on—the necessary medications.

Recently, a CRISPR system was used to eradicate HIV-1 from a humanized mouse model of the disease (Binnie et al., 2021). Antiretroviral therapy combined with CRISPR gene-editing resulted in the elimination of HIV-1, representing a potential amelioration of the virus that causes AIDS with a single, low-cost treatment.

Other viral infections that are current targets for CRISPR in both diagnostics and treatment are Hepatitis B, Epstein Barr Virus, and Human Papilloma Virus (Binnie et al., 2021). In 2021, findings of a CRISPR study focused on the Hepatitis B virus were published in *Molecular Therapy Methods and Clinical Development* (Stone et al., 2021); findings focused on Epstein Barr Virus were published in *PLoS Pathogens* (Akidil et al., 2021); and findings focused on Human Papilloma Virus were published in *PLoS One* (Jubair et al., 2021). Although these studies were all performed in animal models, this represents the first step toward future trials using CRISPR to address infectious diseases in human subjects.

1.4.9 Protein-Folding Conditions

Protein misfolding is believed to be the primary cause of many degenerative and neurodegenerative conditions. These include Alzheimer's disease, Parkinson's disease, Huntington's Chorea, Creutzfeldt-Jakob disease, Cystic Fibrosis, and Gaucher's disease (Chaudhuri & Paul, 2006). While treatments for some of these disorders exist, no effective treatment exists for most progressive conditions. Work in Alzheimer's disease (AD) provides a clear example of how CRISPR represents treatment potential for neurodegenerative conditions. Expression of a form of the apolipoprotein E gene called APOE4 confers a high risk for developing AD, and the protein apoE4 represents a high potential target for therapeutics (Safieh et al., 2019; Sienski et al., 2021). Converting the deleterious form of apoE4 to the related apoE3 (that represents no AD risk) or even the potentially protective apoE2 may be a way to decrease AD risk factors with aging (Safieh et al., 2019). Human stem cells from individuals with AD can model AD and test for potential therapeutics using gene-editing approaches (Sproul et al., 2014). To this end, Wadhwani et al. (2019) used CRISPR to correct the expression of the *APOE4* allele to *APOE3* in two induced pluripotent stem cell (iPSC) lines taken from two individuals with AD. The correction of *APOE4* to *APOE3* protected neurons derived from iPSCs from cell death related to the expression of the APOE4 allele. This represents how CRISPR can alter gene expression in subtypes of cells associated with the development of neurodegenerative conditions.

Due to increased longevity in developed nations such as the US (and, therefore, causing a skew in age-based demographics toward a larger elderly population), it is anticipated by public health experts that the proportion of the population in the US living with dementia (such as Alzheimer's disease and Lewy Body dementia) will increase substantially over the next 20 years. Family members (as well as healthcare providers, hospitals, and nursing homes) are already overburdened in providing care for people living with dementia—resulting in an increased national cost burden attributable to dementia care. Therefore, developing therapeutics to prevent and/or decrease the progression of Alzheimer's disease and/or Lewy body dementia) continues to be a focus of many clinical research studies.

Since the underlying mechanisms leading to the symptoms exhibited in these conditions differ, finding a cure has remained elusive. However, a screening assay for misfolded proteins (such as those linked to AD) utilizing CRISPR has been developed by a team of researchers at the Massachusetts Institute of Technology (MIT). When combined with appropriate phenotypic readouts, these types of medium and high throughput screening assays represent a powerful toolset for the development of personalized medicine approaches across a large number of modalities (Lunden et al., 2020).

1.5 Vaccine Development Utilizing CRISPR—Focus on Covid-19

The first outbreak of the zoonotic SARS-CoV-2 virus—more commonly called Covid-19—initiated in Wuhan, China and was initially reported to the World Health Organization (WHO) in December 2020. Since then, it has infected at least 184 million people worldwide as of July 2021. The death rate has been exceptionally high among the elderly, people with compromised immune systems, and specific chronic health conditions (obesity, Type 2 diabetes, and Chronic Obstructive Pulmonary Disorder [COPD]). Therefore, developing vaccines to eliminate the potential for infection (thereby increasing the likelihood of attaining herd immunity in the population against this coronavirus) has been a priority for physicians, medical researchers, pharmaceutical/biotechnology companies, and worldwide governments.

CRISPR was a critical tool in successfully developing rapid diagnostic tests for Covid-19, (Straiton, 2020), and gene-editing technology was foundational to the development of the mRNA vaccines that have thus far been over 94% effective against infection. According to the Innovative Genomics Institute, a CRISPR-based enhancer of DNA vaccines for Covid-19 is now in development (Innovative Genomics Institute, 2021). While the Covid-19 pandemic has caused untold misery, sickness, and death, its possible silver lining is that it has ushered in major medical research focus on utilizing CRISPR to develop vaccines against respiratory (and other) viral infections.

1.6 Current and Future Challenges

Other than off-target effects, several current scientific challenges hinder the progress of the application of CRISPR to more systems and conditions. As we alluded to earlier, the first is the precise targeting of CRISPR in cells. Rees and Liu (2018) suggest that only 28% of SNPs can be effectively targeted by base editors that utilize Cas9. This is because, as stated earlier, Cas9 activity requires a PAM in the target DNA sequence, and not all potentially targeted sequences are adjacent to PAM sequences (Walton et al., 2020). There are three current approaches to addressing this challenge. The first is the development of PAM-free nucleases. These would confer the ability to edit any sequence in the genome regardless of the presence of a PAM (Collias & Beisel, 2021). However, this may require tradeoffs in editing efficiency as nucleases would have to scan the entire sequence and increase potential off-target effects (Sternberg et al., 2014). Second is the use and discovery of additional Cas proteins with less stringent PAM sequence requirements. Again, a tradeoff would be the potential for off-target effects. Third, the engineering of current Cas nucleases (like Cas12a) that change the nature of PAM recognition. This has been demonstrated in Cas9 via the development of modifications in the enzyme into Cas9-NRNH and Cas9-SpRY (Miller et al., 2020; Walton et al., 2020). These Cas9 modifications allow the recognition of more editable sequences and provide more gene-editing opportunities, albeit at the cost of loss of target specificity (Fig. 1.4).

Cas9 itself presents a challenge concerning safety, depending on the application of CRISPR. The most common sources of Cas9 are *Streptococcus aureus* and *Streptococcus pyogenes*. Because these are bacteria that commonly infect humans, it has been suggested that Cas9 may induce an immunogenic response (Chew et al., 2016; Zhang, 2020). To this end, Zhang (2020) states:

> A recent study by Charlesworth and colleagues show that there are about 78% and 58% tested people had IgG antibodies in serum against SaCas9 and SpCas9 (Charlesworth et al., 2019). They also found that 78% and 67% of T cells against SaCas9 and SpCas9 in the tested people, respectively. Similar results were also observed in other studies (Ferdosi

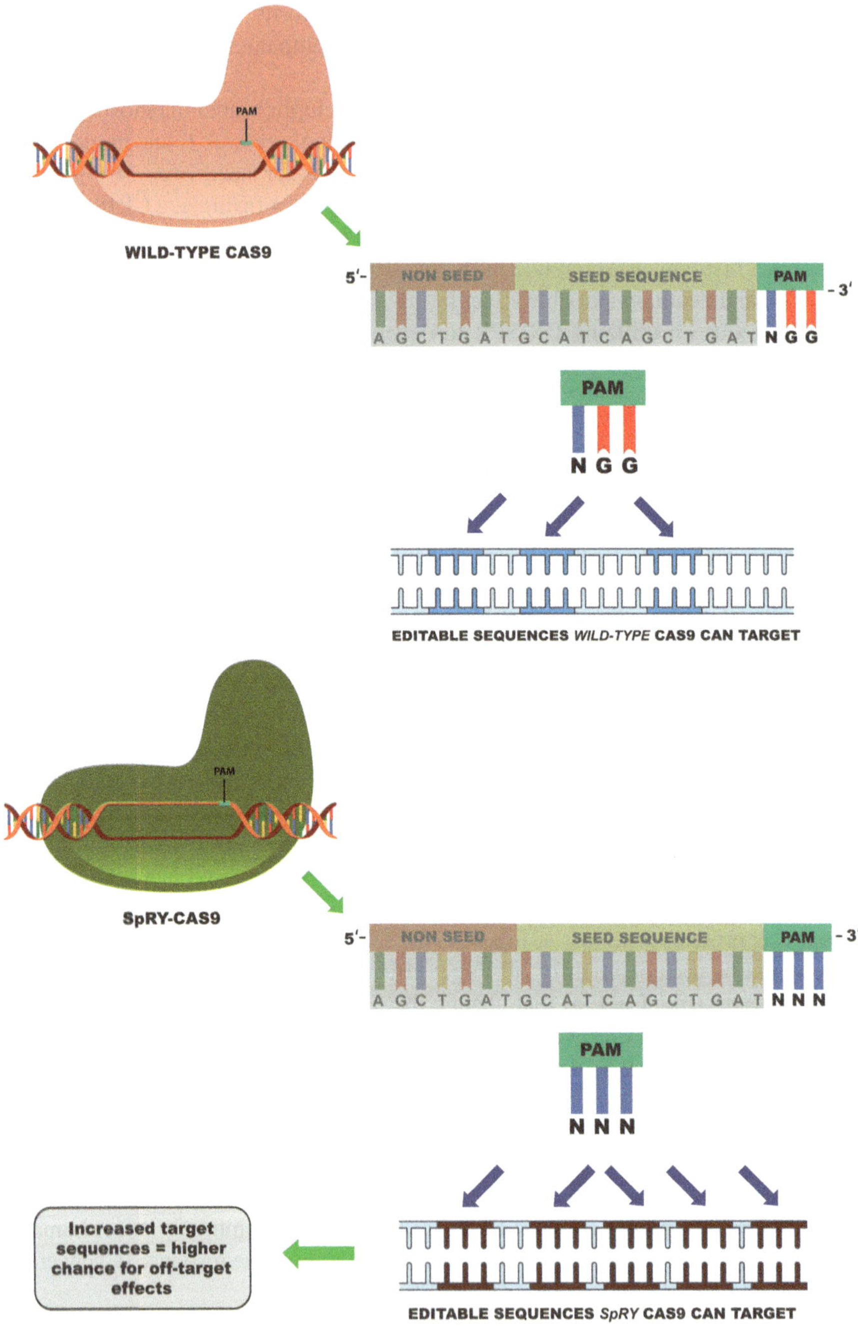

Fig. 1.4 Because of concerns requiring a PAM sequence to engage in CRISPR-editing, which limits the genomic sequences that can be targeted, efforts to develop a PAM-less Cas9 resulted in developing an engineered variant Cas9 from *Streptococcus pyogenes* called Cas9-SpRY. Cas9-SpRY allows for Cas9 access to mutations in the previously inaccessible genome and the ability to edit more sites on the genome. However, the trade-off for this technology is a more significant propensity for off-target effects

et al., 2019; Wagner et al., 2019). All these studies suggest that there are pre-existing humoral and cell-mediated adaptive immune responses to Cas9 in humans (Charlesworth et al., 2019) and that pre-existing humoral and cell-mediated adaptive immune responses to Cas9 in humans may affect the efficiency of CRISPR/Cas9-based gene therapy.

The potential for CRISPR to revolutionize the development of groundbreaking therapeutics is high. But we need to be conscious that large-scale utilization of CRISPR could be akin to opening "Pandora's Box" if we completely embrace CRISPR-gene-editing without any qualms or reservations. Along with the already-suggested potential for off-target impacts (with unknown consequences) in individual patients receiving CRISPR gene-editing, another major challenge posed by widespread embracement of CRISPR is the possibility of its utilization either inadvertently (or intentionally) in a way that leads to unanticipated/unwanted genetic changes in the next-generation of humans and ever-forward. Currently, the possibility of off-target impacts is high due to the lack of complete understanding of the CRISPR machinery and the totality of roles associated with each gene in the human genome.

Even solely in use as therapy to treat a condition, the complex roles of each gene and their complex interaction in humans are not yet fully understood. Therefore, a potential exists that an alteration aimed at treating a condition (such as by boosting specific immune system processes to be more proactive than expected for a given person) could result in the promotion of some other condition in that patient (such as an autoimmune disorder or rare type of cancer—sometimes called *antagonistic pleiotropy*).

A significant challenge for CRISPR (and all gene-editing tools) is linked to the human propensity to want to produce future generations of humans that match our conception of the "best" genetic traits (which fostered such despicable movements as Nazism, leading to genocide and World War II, only 80 years ago).

In 1997, an international legal agreement under the *Council of Europe*'s *Convention on Human Rights and Biomedicine* resulted in a binding treaty that bans inheritable genetic modification and human reproductive cloning (along with *UNESCO* and *World Health Assembly* agreements); however, it was broken by a medical research team in 2018. Biophysicist He Jiankui of China in 2018 announced his research team's creation of the world's first gene-edited babies—albeit to prevent the capacity for HIV infection, as a protective measure—with resulting twin girls born in 2018. Notably, Jiankui was sentenced by a Chinese court to prison for "illegal medical practice" (Cyranoski, 2020). Given the human history of the pursuit of achievements at whatever cost, the aforementioned occurrence is highly unlikely to be the last global attempt at creating gene-edited babies—and despite the prospect of collective revulsion and/or legal retribution. Recent guidelines issued by the World Health Organization (WHO) in the 2021 technical document *Human genome editing: recommendations* attempt to address some of the ethical issues with CRISPR use, and we will discuss this later in the manuscript. As we have used CRISPR in our research, we see the promise as it is used to ameliorate many devastating human conditions. However, there are exceptional cases where we believe that red lines need to be drawn. This manuscript is concerned with developing a framework for understanding ethical challenges surrounding CRISPR use. We contend that international guidelines and registries for CRISPR research like the WHO proposed are a great start, but ultimately not enforceable unless we have international treaties that draw red lines and bind countries to action, following the model that we have used for the proliferation of nuclear weapons. This will need the involvement of agencies like the UN.

Controlling the spread of nuclear weapons has been much harder than nuclear physicists envisioned in the 1950s, and preventing the spread of scientists' impetus to gene edit embryos to conform to a given society's preferences may be no less daunting globally in the decades to come. Ethical decision-making related to CRISPR (and future gene-editing tools) may prove to be more difficult than creating or improving these tools. In the medical research and clinical medical fields, we must always consider the ethics of our activities, and this break-through biotechnology is no exception. Subsequent chapters will develop a framework for keeping pace with this fast-moving, revolutionary gene-editing technology.

References

Akidil, E., Albanese, M., Buschle, A., et al. (2021). Highly efficient CRISPR-Cas9-mediated gene knockout in primary human B cells for functional genetic studies of Epstein-Barr virus infection. *PLoS Pathogens, 17*(4), e1009117. Webpage: https://www.ncbi.nlm.nih.gov/pmc/articles/PMC8078793/

Anzalone, A. V., Randolph, P. B., Davis, J. R., et al. (2019). Search-and-replace genome editing without double-strand breaks or donor DNA. *Nature, 576*, 149–157. https://doi.org/10.1038/s41586-019-1711-4

Bhamidipati, K., Silberstein, J. L., Chaichian, Y., Baker, M. C., Lanz, T. V., Zia, A., … Robinson, W. H. (2021). CD52 is elevated on B cells of SLE patients and regulates B cell function. *Frontiers in Immunology, 11*. https://doi.org/10.3389/fimmu.2020.626820

Binnie, A., Fernandes, E., Almeida-Lousada, H., Mello, R. A., & Castelo-Branco, P. (2021). CRISPR-based strategies in infectious disease diagnosis and therapy. *Infection, 49*(3), 377–385. https://doi.org/10.1007/s15010-020-01554-w

Broad Institute. Home. CRISPR – CRISPR Timeline. (2016). Webpage: https://www.broadinstitute.org/what-broad/areas-focus/project-spotlight/crispr-timeline

Bulcha, J. T., Wang, Y., Ma, H., Tai, P. W., & Gao, G. (2021). Viral vector platforms within the gene therapy landscape. *Signal Transduction and Targeted Therapy, 6*(1). https://doi.org/10.1038/s41392-021-00487-6

Canalis, E., Yu, J., Schilling, L., Yee, S., & Zanotti, S. (2018). The lateral meningocele syndrome mutation causes marked osteopenia in mice. *Journal of Biological Chemistry, 293*(36), 14165–14177. https://doi.org/10.1074/jbc.ra118.004242

Charlesworth, C. T., Deshpande, P. S., Dever, D. P., Camarena, J., Lemgart, V. T., Cromer, M. K., … Porteus, M. H. (2019). Identification of preexisting adaptive immunity to Cas9 proteins in humans. *Nature Medicine, 25*(2), 249–254. https://doi.org/10.1038/s41591-018-0326-x

Chaudhuri, T. K., & Paul, S. (2006). Protein-misfolding diseases and chaperone-based therapeutic approaches. *FEBS Journal, 273*(7), 1331–1349. https://doi.org/10.1111/j.1742-4658.2006.05181.x

Chen, Y. C., Farzadfard, F., Gharaei, N., et al. (2017). Randomized CRISPR-Cas transcriptional perturbation screening reveals protective genes against alpha-synuclein toxicity. *Molecular Cell, 68*(1), 247–257.e5. Webpage: https://pubmed.ncbi.nlm.nih.gov/28985507/

Chen, J. S., Ma, E., Harrington, L. B., et al. (2018). CRISPR-Cas12a target binding unleashes indiscriminate single-stranded DNase activity. *Science, 360*, 436–439. Webpage: https://www.ncbi.nlm.nih.gov/pmc/articles/PMC6628903/

Chew, W. L., Tabebordbar, M., Cheng, J. K., Mali, P., Wu, E. Y., Ng, A. H., … Church, G. M. (2016). A multifunctional AAV–CRISPR–Cas9 and its host response. *Nature Methods, 13*(10), 868–874. https://doi.org/10.1038/nmeth.3993

Cohen, J. (2017, February 15). How the battle lines over CRISPR were drawn. *Science*. Webpage: https://www.sciencemag.org/news/2017/02/how-battle-lines-over-crispr-were-drawn

Collias, D., & Beisel, C. L. (2021). CRISPR technologies and the search for the PAM-free nuclease. *Nature Communications, 12*(1). https://doi.org/10.1038/s41467-020-20633-y

Cullot, G., Boutin, J., Toutain, J., Prat, F., Pennamen, P., Rooryck, C., … Bedel, A. (2019). CRISPR-Cas9 genome editing induces megabase-scale chromosomal truncations. *Nature Communications, 10*(1). https://doi.org/10.1038/s41467-019-09006-2

Cyranoski, D. (2020, January 3). What CRISPR-baby prison sentences mean for research. *Nature*. Webpage: https://www.nature.com/articles/d41586-020-00001-y

Czarnek, M., & Bereta, J. (2016). The CRISPR-Cas system – From bacterial immunity to genome engineering. *Postępy Higieny i Medycyny Doświadczalnej* (Online), *70*(0), 901–916. Webpage: https://pubmed.ncbi.nlm.nih.gov/27594566/

Durens, M., Nestor, J., Williams, M., Herold, K., Niescier, R. F., Lunden, J. W., … Nestor, M. W. (2020). High-throughput screening of human induced pluripotent stem cell-derived brain organoids. *Journal of Neuroscience Methods, 335*, 108627. https://doi.org/10.1016/j.jneumeth.2020.108627

Enache, O. M., Rendo, V., Abdusamad, M., Lam, D., Davison, D., Pal, S., … Ben-David, U. (2020). Cas9 activates the p53 pathway and selects for p53-inactivating mutations. *Nature Genetics, 52*(7), 662–668. https://doi.org/10.1038/s41588-020-0623-4

Feehan, J. M., Chiu, C. N., Stanar, P., Tam, B. M., Ahmed, S. N., & Moritz, O. L. (2017). Modeling dominant and recessive forms of retinitis pigmentosa by editing three rhodopsin-encoding genes in Xenopus laevis using Crispr/Cas9. *Scientific Reports, 7*(1). https://doi.org/10.1038/s41598-017-07153-4

Ferdosi, S. R., Ewaisha, R., Moghadam, F., Krishna, S., Park, J. G., Ebrahimkhani, M. R., … Anderson, K. S. (2019). Multifunctional CRISPR-Cas9 with engineered immunosilenced human T cell epitopes. *Nature Communications, 10*(1). https://doi.org/10.1038/s41467-019-09693-x

Gao, Y., Gao, K., & Yang, H. (2020). CRISPR/Cas: A potential gene-editing tool in the nervous system. *Cell Regeneration, 9*, 12. https://doi.org/10.1186/s13619-020-00044-6

Gasiunas, G., Barrangou, R., Horvath, P., & Siksnys, V. (2012). Cas9-crRNA ribonucleoprotein complex mediates specific DNA cleavage for adaptive immunity in bacteria. *Proceedings of the National Academy of Sciences, 109*(39). https://doi.org/10.1073/pnas.1208507109

Gaudelli, N. M., Komor, A. C., Rees, H. A., Packer, M. S., Badran, A. H., Bryson, D. I., & Liu, D. R. (2018). Publisher correction: Programmable base editing of A•T to G•C in genomic DNA without DNA cleavage. *Nature, 559*(7714). https://doi.org/10.1038/s41586-018-0070-x

Geurts, M. H., Poel, E. D., Amatngalim, G. D., Oka, R., Meijers, F. M., Kruisselbrink, E., … Clevers, H. (2020). CRISPR-based adenine editors correct nonsense mutations in a cystic fibrosis organoid biobank. *Cell Stem Cell, 26*(4). https://doi.org/10.1016/j.stem.2020.01.019

Gootenberg, J. S., Abudayyeh, O. O., Lee, J. W., et al. (2017). Nucleic acid detection with CRISPR-Cas13a/C2c2. *Science, 356*(6336), 438–442. Webpage: https://www.ncbi.nlm.nih.gov/pmc/articles/PMC5526198/

Grünewald, J., Zhou, R., Lareau, C. A., Garcia, S. P., Iyer, S., Miller, B. R., … Joung, J. K. (2020). A dual-deaminase CRISPR base editor enables concurrent adenine and cytosine editing. *Nature Biotechnology, 38*(7), 861–864. https://doi.org/10.1038/s41587-020-0535-y

Guo, X., Chavez, A., Tung, A., Chan, Y., Kaas, C., Yin, Y., … Church, G. M. (2018). High-throughput creation and functional profiling of DNA sequence variant libraries using CRISPR–Cas9 in yeast. *Nature Biotechnology, 36*(6), 540–546. https://doi.org/10.1038/nbt.4147

Ha, T. K., Hansen, A. H., Kildegaard, H. F., & Lee, G. M. (2020). Knockout of sialidase and pro-apoptotic genes in Chinese hamster ovary cells enables the production of recombinant human erythropoietin in fed-batch cultures. *Metabolic Engineering, 57*, 182–192. https://doi.org/10.1016/j.ymben.2019.11.008

Henderson, H. (2021, March 3). CRISPR clinical trials: A 2021 update. Innovative Genomics Institute. Webpage: https://innovativegenomics.org/news/crispr-clinical-trials-2021/

Innovative Genomics Institute. (2021) CRISPR-based enhancers of DNA vaccines for Covid-19. Webpage: https://innovativegenomics.org/projects/crispr-based-dna-vaccine-enhancer-covid-19/

Ishino, Y., Krupovic, M., & Forterre, P. (2018). History of CRISPR-Cas from encounter with a mysterious repeated sequence to genome editing technology. *Journal of Bacteriology, 200*(7), e00580-17. Webpage: https://www.ncbi.nlm.nih.gov/pmc/articles/PMC5847661/

Jain, S., Shukla, S., Yang, C., Zhang, M., Fatma, Z., Lingamaneni, M., … Zhao, H. (2021). TALEN outperforms Cas9 in editing heterochromatin target sites. *Nature Communications, 12*(1). https://doi.org/10.1038/s41467-020-20672-5

Jakobsen, R. K. (2020, April 28). First Chinese CRISPR gene therapy trial demonstrates safety. *CRISPR Medicine News*. Webpage: https://crisprmedicinenews.com/news/first-chinese-crispr-gene-therapy-trial-demonstrates-safety/

Jansen, R., Embden, J. D., Gaastra, W., et al. (2002). Identification of genes that are associated with DNA repeats in prokaryotes. *Molecular Microbiology, 43*(6), 1565–1575. Webpage: https://pubmed.ncbi.nlm.nih.gov/11952905/

Jinek, M., Chylinski, K., Fonfara, I., Hauer, M., Doudna, J. A., & Charpentier, E. (2012). A programmable dual-RNA-guided DNA endonuclease in adaptive bacterial immunity. *Science, 337*(6096), 816–821. https://doi.org/10.1126/science.1225829

Jubair, L., Lam, A. K., Fallaha, S., et al. (2021). CRISPR/Cas9-loaded stealth liposomes effectively cleared established HPV16-driven tumours in syngeneic mice. *PLoS One, 16*(1), e0223288. Webpage: https://journals.plos.org/plosone/article?id=10.1371/journal.pone.0223288

Kaiser, J. (2021, June 26). CRISPR injected into the blood treats a genetic disease for first time. *Science*. Webpage: https://www.sciencemag.org/news/2021/06/crispr-injected-blood-treats-genetic-disease-first-time

Khan, S., Mahmood, M. S., Rahman, S. U., et al. (2018a). CRISPR/Cas9: The Jedi against the dark empire of diseases. *Journal of Biomedical Science, 25*(1), 29. Webpage: https://www.ncbi.nlm.nih.gov/pmc/articles/PMC5875004/

Khan, M. Z., Amin, I., Hameed, A., & Mansoor, S. (2018b). CRISPR–Cas13a: Prospects for plant virus resistance. *Trends in Biotechnology, 36*(12), 1207–1210. https://doi.org/10.1016/j.tibtech.2018.05.005

Kim, J., Koo, B., & Knoblich, J. A. (2020). Human organoids: Model systems for human biology and medicine. *Nature Reviews Molecular Cell Biology, 21*(10), 571–584. https://doi.org/10.1038/s41580-020-0259-3

Kitamoto, K., Taketani, Y., Fujii, W., Inamochi, A., Toyono, T., Miyai, T., … Ouchi, Y. (2020). Generation of mouse model of TGFBI-R124C corneal dystrophy using CRISPR/Cas9-mediated homology-directed repair. *Scientific Reports, 10*(1). https://doi.org/10.1038/s41598-020-58876-w

Komor, A. C., Kim, Y. B., Packer, M. S., Zuris, J. A., & Liu, D. R. (2016). Programmable editing of a target base in genomic DNA without double-stranded DNA cleavage. *Nature, 533*(7603), 420–424. https://doi.org/10.1038/nature17946

Lander, E. (2016). The heroes of CRISPR. *Cell, 164*(1–2), 18–28. https://doi.org/10.1016/j.cell.2015.12.041

Ledford, H. (2018, July 16). CRISPR gene editing produces unwanted DNA deletions. Nature News. Retrieved April 18, 2022, from https://www.nature.com/articles/d41586-018-05736-3

Ledford, H. (2020, March 5). News – CRISPR treatment inserted directly into the body for first time. *Nature*. Webpage: https://www.nature.com/articles/d41586–020-00655-8

Liu, J., Zhou, G., Zhang, L., & Zhao, Q. (2019). Building potent chimeric antigen receptor T cells with CRISPR genome editing. *Frontiers in Immunology, 10*. https://doi.org/10.3389/fimmu.2019.00456

Lunden, J. W., Durens, M., Nestor, J., Niescier, R. F., Herold, K., Brandenburg, C., … Nestor, M. W. (2020). Development of a 3-D organoid system using human induced pluripotent stem cells to model idiopathic autism. *Advances in Neurobiology Neurodevelopmental Disorders*, 259–297. https://doi.org/10.1007/978-3-030-45493-7_10

Maddalo, D., Manchado, E., Concepcion, C. P., Bonetti, C., Vidigal, J. A., Han, Y., … Ventura, A. (2014). In vivo engineering of oncogenic chromosomal rearrangements with the CRISPR/Cas9 system. *Nature, 516*(7531), 423–427. https://doi.org/10.1038/nature13902

Makarova, K. S., Wolf, Y. I., Iranzo, J., Shmakov, S. A., Alkhnbashi, O. S., Brouns, S. J., … Koonin, E. V. (2019). Evolutionary classification of CRISPR–Cas systems: A burst of class 2 and derived variants. *Nature Reviews Microbiology, 18*(2), 67–83. https://doi.org/10.1038/s41579-019-0299-x

Matano, M., Date, S., Shimokawa, M., Takano, A., Fujii, M., Ohta, Y., … Sato, T. (2015). Modeling colorectal cancer using CRISPR-Cas9–mediated engineering of human intestinal organoids. *Nature Medicine, 21*(3), 256–262. https://doi.org/10.1038/nm.3802

Miller, S. M., Wang, T., Randolph, P. B., Arbab, M., Shen, M. W., Huang, T. P., … Liu, D. R. (2020). Continuous evolution of SpCas9 variants compatible with non-G PAMs. *Nature Biotechnology, 38*(4), 471–481. https://doi.org/10.1038/s41587-020-0412-8

Min, Y., Bassel-Duby, R., & Olson, E. N. (2019). CRISPR correction of Duchenne muscular dystrophy. *Annual Review of Medicine, 70*(1), 239–255. https://doi.org/10.1146/annurev-med-081117-010451

Nguyen, T. H., & Anegon, I. (2016). Successful correction of hemophilia by CRISPR /Cas9 genome editing in vivo: Delivery vector and immune responses are the key to success. *EMBO Molecular Medicine, 8*(5), 439–441. https://doi.org/10.15252/emmm.201606325

Palermo, G., Ricci, C. G., Fernando, A., Basak, R., Jinek, M., Rivalta, I., … Mccammon, J. A. (2017). Protospacer adjacent motif-induced allostery activates CRISPR-Cas9. *Journal of the American Chemical Society, 139*(45), 16028–16031. https://doi.org/10.1021/jacs.7b05313

Pausch, P., Al-Shayeb, B., Bisom-Rapp, E., Tsuchida, C. A., Li, Z., Cress, B. F., … Doudna, J. A. (2020). CRISPR-CasΦ from huge phages is a hypercompact genome editor. *Science, 369*(6501), 333–337. https://doi.org/10.1126/science.abb1400

Rees, H. A., & Liu, D. R. (2018). Base editing: Precision chemistry on the genome and transcriptome of living cells. *Nature Reviews Genetics, 19*(12), 770–788. https://doi.org/10.1038/s41576-018-0059-1

Richardson, C. D., Kazane, K. R., Feng, S. J., Zelin, E., Bray, N. L., Schäfer, A. J., … Corn, J. E. (2018). CRISPR–Cas9 genome editing in human cells occurs via the Fanconi anemia pathway. *Nature Genetics, 50*(8), 1132–1139. https://doi.org/10.1038/s41588-018-0174-0

Rivera-Torres, N., Banas, K., Bialk, P., Bloh, K. M., & Kmiec, E. B. (2017). Insertional mutagenesis by CRISPR/Cas9 ribonucleoprotein gene editing in cells targeted for point mutation repair directed by short single-stranded DNA oligonucleotides. *PLoS One, 12*(1). https://doi.org/10.1371/journal.pone.0169350

Ryu, S., Koo, T., Kim, K., Lim, K., Baek, G., Kim, S., … Kim, J. (2018). Adenine base editing in mouse embryos and an adult mouse model of Duchenne muscular dystrophy. *Nature Biotechnology, 36*(6), 536–539. https://doi.org/10.1038/nbt.4148

Safieh, M., Korczyn, A. D., & Michaelson, D. M. (2019). ApoE4: An emerging therapeutic target for Alzheimer's disease. *BMC Medicine, 17*(1), 64. https://doi.org/10.1186/s12916-019-1299-4. PMID: 30890171; PMCID: PMC6425600.

Sakata, R. C., Ishiguro, S., Mori, H., Tanaka, M., Tatsuno, K., Ueda, H., Yamamoto, S., Seki, M., Masuyama, N., Nishida, K., Nishimasu, H., Arakawa, K., Kondo, A., Nureki, O., Tomita, M., Aburatani, H., & Yachie, N. (2020). Base editors for simultaneous introduction of C-to-T and A-to-G mutations. *Nature Biotechnology, 38*(7), 865–869. https://doi.org/10.1038/s41587-020-0509-0. Epub 2020 Jun 2.

Sansbury, B. M., Hewes, A. M., & Kmiec, E. B. (2019). Understanding the diversity of genetic outcomes from CRISPR-Cas generated homology-directed repair. *Communications Biology, 2*, 458 (article 458; Open Access). Webpage: https://www.nature.com/articles/s42003-019-0705-y

Schwank, G., Koo, B., Sasselli, V., Dekkers, J., Heo, I., Demircan, T., … Clevers, H. (2013). Functional repair of CFTR by CRISPR/Cas9 in intestinal stem cell organoids of cystic fibrosis patients. *Cell Stem Cell, 13*(6), 653–658. https://doi.org/10.1016/j.stem.2013.11.002

Shmakov, S. A., Makarova, K. S., Wolf, Y. I., Severinov, K. V., & Koonin, E. V. (2018). Systematic prediction of genes functionally linked to CRISPR-Cas systems by gene neighborhood analysis. *Proceedings of the National Academy of Sciences, 115*(23). https://doi.org/10.1073/pnas.1803440115

Sienski, G., et al. (2021). APOE4 disrupts intracellular lipid homeostasis in human iPSC-derived glia. *Science Translational Medicine, 13*(583), eaaz4564. https://doi.org/10.1126/scitranslmed.aaz4564

Sproul, A. A., Jacob, S., Pre, D., Kim, S. H., Nestor, M. W., Navarro-Sobrino, M., …Noggle, S. A. (2014). Characterization and molecular profiling of PSEN1 familial alzheimer's disease IPSC-derived neural progenitors. *PLoS ONE, 9*(1). https://doi.org/10.1371/journal.pone.0084547

Sternberg, S. H., Redding, S., Jinek, M., Greene, E. C., & Doudna, J. A. (2014). DNA interrogation by the CRISPR RNA-guided endonuclease Cas9. *Nature, 507*(7490), 62–67. https://doi.org/10.1038/nature13011

Stone, D., Long, K. R., Loprieno, M. A., et al. (2021). CRISPR-Cas9 gene editing of hepatitis B virus in chronically infected humanized mice. *Molecular Therapy Methods and Clinical Development, 20*, 258–275. Webpage: https://www.sciencedirect.com/science/article/pii/S2329050120302412

Straiton, J. (2020). CRISPR vs COVID-19: How can gene editing help beat a virus? *BioTechniques, 69*(5) (Open Access). Webpage: https://www.future-science.com/doi/10.2144/btn-2020-0145

Vidyasagar, A. (2018, April 20). What is CRISPR? *LiveScience.com*. Webpage: https://www.livescience.com/58790-crispr-explained.html

Wadhwani, A. R., Affaneh, A., Gulden, S. V., & Kessler, J. A. (2019). Neuronal apolipoprotein E4 increases cell death and phosphorylated tau release in Alzheimer disease. *Annals of Neurology, 85*(5), 726–739. https://doi.org/10.1002/ana.25455

Wagner, D. L., Amini, L., Wendering, D. J., Burkhardt, L. M., Akyüz, L., Reinke, P., Volk, H. D., & Schmueck-Henneresse, M. (2019). High prevalence of Streptococcus pyogenes Cas9-reactive T cells within the adult human population. *Nature Medicine, 25*(2), 242–248. https://doi.org/10.1038/s41591-018-0204-6. Epub 2018 Oct 29. PMID: 30374197.

Walton, R. T., Christie, K. A., Whittaker, M. N., & Kleinstiver, B. P. (2020). Unconstrained genome targeting with near-PAMless engineered CRISPR-Cas9 variants. *Science, 368*(6488), 290–296. https://doi.org/10.1126/science.aba8853

Yan, S., Tu, Z., Liu, Z., Fan, N., Yang, H., Yang, S., … Li, X. (2018). A huntingtin knockin pig model recapitulates features of selective neurodegeneration in Huntington's disease. *Cell, 173*(4). https://doi.org/10.1016/j.cell.2018.03.005

Yip, B. H. (2020). Recent advances in CRISPR/Cas9 delivery strategies. *Biomolecules, 10*(6), 839. https://doi.org/10.3390/biom10060839

Yoshihara, E., O'Connor, C., Gasser, E., Wei, Z., Oh, T. G., Tseng, T. W., … Evans, R. M. (2020). Immune-evasive human islet-like organoids ameliorate diabetes. *Nature, 586*(7830), 606–611. https://doi.org/10.1038/s41586-020-2631-z

Zamani, M., Robson, J. M., Fan, A., Bono, M. S., Furst, A. L., & Klapperich, C. M. (2021). Electrochemical strategy for low-cost viral detection. *ACS Central Science, 7*(6), 963–972. https://doi.org/10.1021/acscentsci.1c00186

Zhang, B. (2020). CRISPR/Cas gene therapy. *Journal of Cellular Physiology, 236*(4), 2459–2481. https://doi.org/10.1002/jcp.30064

Zmajkovic, J., Lundberg, P., Nienhold, R., Torgersen, M. L., Sundan, A., Waage, A., & Skoda, R. C. (2018). A gain-of-function mutation inEPOin familial erythrocytosis. *New England Journal of Medicine, 378*(10), 924–930. https://doi.org/10.1056/nejmoa1709064

Chapter 2
Introduction to Anticipatory Ethics

2.1 Introduction

Anticipatory Ethics has recently emerged as an important new orientation in practical ethics (Brey, 2012a; Johnson, 2010; Nestor et al., 2014; Nestor & Wilson, 2019; Verbeek, 2005, 2011). This form of practical ethics is paramount to keeping ethical analyses at pace with the rapid development of new technologies. An anticipatory ethical analysis converges a multitude of approaches onto stakeholder analyses, employing modern computational and statistical techniques with traditional ethical frameworks such as consequentialism, utilitarianism, and intuitive ethics. The synergy between stakeholder analysis buttressed by a supporting ethical framework allows the user to approximate an aggregate feedback loop that involves both the users and developers of advanced and emergent technologies. Examples of this approach are the use of questionnaires, interviews, nominal group technique, and Delphi panels to elucidate and quantify stakeholder opinions and feedback. Feedback obtained using these and other subjective analyses can be combined with participatory foresight and other forecasting models to develop models of future lived experiences that stakeholders can build consensus around (Nikolova, 2013). Generated future-facing models can either be subjected to ethical analysis at the beginning, end, or throughout the entire anticipatory process. Often, these models are created developing scenarios that Nanayakkara et al. (2020) describe as:

> …predictive ("what will happen?"), explorative ("what could happen?") and normative ("how can a certain target be reached?")

Anticipatory ethics in the context of emerging advanced technology is focused on both the predictive and explorative aspects of the technology, emphasizing the explorative modality. The data and models generated using an anticipatory approach and with a proper practical ethical analysis can be used to inform policy and other decision-makers within the context of anticipatory governance:

M. W. Nestor, R. L. Wilson, *Anticipatory Ethics and The Use of CRISPR in Humans*, https://doi.org/10.1007/978-3-030-98368-0_2

> Anticipatory governance is typically concerned with the prediction of future behaviors of individuals by collecting large amounts of data and making forecasted policy based on predictions necessitated by the data. These two types of analysis (anticipatory ethics and anticipatory governance) are readily combined into anticipatory ethics and governance, which makes policy based on data that engages the public and tackles a future-oriented legal analysis based on theories of justice for rapidly emerging technologies (Hester et al., 2015; Nestor & Wilson, 2020).

In this manuscript, an anticipatory ethical analysis will be focused on explorative ideas centered around the potential future use of CRISPR gene-editing technology in humans. Although CRISPR technology has a growing number of applications outside its use in humans (examples include: increasing insect resistance in crops or susceptibility to pesticides in mosquitoes), it is essential to focus our discussion on a narrow set of cases concerning human use. It is also essential to underscore that many applications of CRISPR in humans may pose little ethical challenge. An example where there may be very little use for an anticipatory ethical analysis is the use of CRISPR to treat sickle cell anemia, for which the first person was treated in 2020 (Demirci et al., 2019; Plater, 2020).

We derive in this manuscript an anticipatory practical ethics based on foundations laid by Moor, Brey, and Johnson, amongst others. A foundational principle running through the analysis in this manuscript is that CRISPR is a revolutionary technology. It will continue to be needed as an essential research tool in many areas of science (Wang et al., 2016). The other thematic thread underlying our analysis is that:

> …CRISPR allows deeper exploration of the relationships between genetics and cell/molecular biology, which will help us to understand the etiology of disease. In doing so, new treatments for disease with potential for the eradication of some diseases will be possible. As such, CRISPR is seen as a net positive for humans. (Nestor & Wilson, 2020)

As we have argued in previous work, this analysis is sound for somatic gene editing, where we see no major ethical dilemma. However, in the context of germline gene editing using CRISPR, we must think through anticipatory ethical challenges and develop policies, as this has to be considered a serious application of CRISPR (Regalado, 2021). We recently wrote:

> As with every revolutionary scientific discovery and how it relates to society at large, there is the risk of unintended uses or consequences for future persons. Thus, we use an anticipatory ethical framework to explore the potential net negative effects for future persons with respect to the use of CRISPR… (Nestor & Wilson, 2020)

The focus of our discussion in this manuscript will be on the use of CRISPR to alter brain function or cognition in humans. We have chosen this aspect to focus on because it provides the most fruitful perspectives for discussion of the ethical implications for using this technology and drives the notion of the *lived experience* and many qualities that we associate with the essence of what it means to be human.

2.2 Some Assumptions

For this analysis, it is crucial to set up some assumptions and definitions at the forefront. First, we will focus on a very narrowly bounded set of stakeholders. For this analysis, we will focus on three groups of stakeholders: scientific researchers, clinicians, and patients. Of course, the web of stakeholders that can be employed in a thorough anticipatory ethical analysis is exceedingly complex and much more extensive, especially concerning medicine and health care.

Because the goal of this manuscript is to develop a *framework* for the development of a more sophisticated anticipatory ethical analysis of the use of CRISPR to alter genes that underlie the function of the human mind, we felt it essential to keep the simple set of stakeholders within the three major and generalized categories we have outlined. This is to focus the discussion and give some simplicity to the analysis. Second, we will colloquially refer to the use of CRISPR in human embryos (germline editing) with the express purpose of enhancing or otherwise modifying the brain simply as "CRISPR" in most contexts. When we are using the term CRISPR, the reader should make this assumption unless explicitly stated otherwise. Finally, we will use the word "condition" instead of the commonly used "disorder." This is a deliberate rhetorical device to remove the emotional response that may occur when describing a genetically inherited phenotype over which one has no control as a "disorder." The term disorder can imply the idea of a "less than" or "abnormal" existence. We do not wish to suggest that any person who inherits a genetic phenotype is inferior. We will posit later in this manuscript that this can be an underlying implicit assumption of some of the stakeholders pushing for the use of CRISPR to ameliorate some types of human conditions.

2.3 A Brief Overview

The importance of a sound anticipatory ethical analysis as a basis for practical ethics has been recognized by several influential philosophers as critical to keeping pace with technology's development and influence in all its forms (Pimple, 2014). As stated earlier, this is because anticipatory ethics focuses on identifying ethical issues surrounding developing technologies and then strategies for identifying and resolving potential ethical problems with these technologies before and as they arise. (Brey, 2012b; Johnson, 2010). Thus, a brief overview of some of the influential philosophers, concepts, and ethical frameworks that have given rise to this new area of ethics may be helpful to readers unfamiliar with anticipatory ethics.

This discussion will focus on three versions of anticipatory ethics. The overview presented here will allow for a survey of the definitions, conceptions, and the focus of different versions of anticipatory ethics. The three forms of anticipatory ethics and the insights developed by the authors of these variations are all related to the general development of technology and the design, development, and use of technological artifacts. From a historical perspective, many of the difficulties in adequately building a sound ethical analysis around developing technologies have their roots in the evolution of philosophical systems and ethical analyses examining the development of technology and the artifacts that exemplify the technology. Because of the way ethicists treat technologies in ethical studies, ambiguities related to what it means to be an "agent" and what it means to "act" have hampered the development of a clear roadmap that can be used to apply classical philosophical ethics to quickly evolving and cutting-edge innovative and disruptive technologies.

It is around these ambiguities (namely, what is an "agent" and what is an "act" during technological development) that the field of anticipatory ethics has evolved. To illustrate this evolution, it is crucial to sketch four representative views of how the relationship between "subjects" and "artifacts" are identified and how four different authors present differing views of anticipatory ethics based upon how subjects and objects are related. A common theme within these three perspectives is the idea that is that anticipatory ethics has to be developed in *conjunction* with "an empirically oriented philosophy of technology" (Achterhuis, 2001) and that this empirical orientation for anticipatory ethics must be in turn anchored in case studies of specific technologies and technological artifacts. The reason for this emphasis on particular technologies and artifacts is due to the exponential rate of development and growth of technology. For example, gene-editing technologies such as CRISPR are undergoing rapid transformation. While each iteration (variation) of the technology is still being developed, each new iteration presents new possibilities that can alter the trajectory of the technology in general. The rapid prototyping and iterative nature of CRISPR in parallel with its use in clinical populations require a complete understanding of stakeholder needs and interests via case studies. Although this is necessary for developing an anticipatory scheme that keeps pace, it is not sufficient. Case studies are rooted in the present and can project into the future. In contrast, exploratory models of stakeholders' future lived experiences are more aligned in the aggregate with the anticipatory ethics/anticipatory governance model. This means that an accurate analysis of CRISPR requires combining both case studies and a forecast model of predicted future lived experiences.

James Moor has forwarded an early framework for developing ethics for emerging technologies. In a piece entitled, "Why we need better ethics for emerging technologies," Moor posits the hypothesis that as technological revolutions progress, ethical problems increase, which he refers to as "Moor's Law":

> Moor's Law: As technological revolutions increase their social impact, ethical problems increase. (Moor, 2005)

Central to Moor's view is the idea that new technological products are constantly developing, and technologies are rapidly expanding. This perpetual development and expansion process creates an increasing set of ethical dilemmas that are not being resolved before new technologies (and accompanying ethical quandaries) arise. Moor isolates and identifies the convergence of genetics and neuroscience as an area in which complex ethical quandaries will develop over time:

> Although genetic technology, nanotechnology, and neurotechnology are not yet fully developed in this regard, they have two features that suggest that such development is likely.

> First, each is driven by a conception of a general capability of the field, a malleability. Just as computing is based on logic malleability, genetic technology is based on life malleability, nanotechnology is based on material malleability, and neurotechnology is based on mental malleability. They offer us the capabilities of building new bodies, new environments, and even new minds. Such fundamental capabilities are very likely to be funded, to be developed, and to have significant social impact. Second, the emerging technologies are converging. They enable each other as tools, as components, and as models. This convergence will move all these technologies forward in a revolutionary path. Thus, we can expect an increase in ethical issues in the future as the technologies mature. (Moor, 2005)

This technological feedback loop requires a redefinition of how practical ethics should be practiced. He writes:

> The ethical issues that we will confront will not only come in increasing numbers but will come packaged in terms of complex technology. Such ethical issues will require considerable effort to be understood as well as a considerable effort to formulate and justify good ethical policies. This will not be ethics as usual. People who both understand the technologies and are knowledgeable about ethics are in short supply just as the need is expanding. (Moor, 2005)

Moor suggests three significant changes to applying traditional ethical frameworks to rapidly emerging technologies. First, he requires a re-definition of ethics from a static and slow-moving (primarily academic) enterprise to one that reframes ethical analysis as dynamic and rapidly iterative. Second, Moor implores the ethics community (heavily biased towards philosophers) to forge more multidisciplinary cross-collaborative relationships with other stakeholders such as scientists, engineers, computer technicians, and any group that can contribute to the conversation developing ethical frameworks for technological analysis. Third, Moor hypothesizes that more sophisticated ethical analyses, presumably taking advantage of new computing technologies to quantify subjective experiences and other complex social science methods, should be employed to create a framework that is in step with dynamic and rapidly evolving cutting-edge technologies.

To this end, Moor reframes the lay definition of technology as an isolated phenomenon to a "technological paradigm," a holistic wrapper that includes the theories, methods, and concepts that contextualize and characterize a technology. Each technology then becomes a specific manifestation of the technological paradigm from which it flows. Technological paradigms lead to specific technologies and technological revolutions, and it is around these technological revolutions and their ethical complexity that Moor builds a foundational framework for an anticipatory ethical analysis. Moor describes a technological revolution as occurring when technological development has reached a stage where this development has had an enormous social impact. According to Moor, technological revolutions occur in three major stages: the introduction stage, the permeation stage, and the power stage. The most critical time for an anticipatory ethical analysis is during the induction stage, an extension of the "pre-revolutionary" period when the technology is just beginning to come into its own. Moor outlines a set of criteria for critical ethical analysis across the three stages of a technological revolution (defined as introduction, permeation, and power). These categories can be considered an early and primary set of criteria for developing a stakeholder and anticipatory ethical technology analysis. These include:

- Devices
- Users/beneficiaries
- Understanding
- Cost per use
- Usefulness
- Integration into society
- Social impact

Each of these criteria can be evaluated and re-evaluated at each point in the technological revolution. The last criterion, social impact, is the most important in an ethical analysis of technology in general, and CRISPR specifically. This is because social impact is the culmination of all the other categories. Moor states:

> Social impact inevitably reflects the other factors mentioned…and in addition includes the effect the technology has on the behavior and practices of the society. A technological revolution has a large scale transforming effect on the manner in which a society functions. (Moor, 2005)

We and others contend that CRISPR as a technological paradigm is inducing a technological revolution in biomedical science. Moor's early framework can be used to contextualize and provide a benchmark for the ethical analysis of this ongoing revolution.

Phillip Brey presents a second framework critical in the development of anticipatory ethics in his article "Anticipating ethical issues in emerging IT." (Brey, 2012b) According to Brey:

> The ethics of emerging technology is the study of ethical issues at the R&D and introduction stage of technology development through the anticipation of possible future devices, applications and social consequences. (Brey, 2012b)

Brey named this ethics "anticipatory technology ethics" (a precursor to 'anticipatory ethics' and ATE for short), and he posits that ATE is essential for analyzing technologies before they enter into Moor's "introductory" stage of development. Brey calls this stage the "research & development stage" (R&D) and writes:

> This stage comes before the introduction stage. Whereas in the introduction stage, there are already early applications of a new technology, in the R&D stage these applications do not yet exist. Rather, research is directed at the development of basic techniques that may further down the road result in concrete applications. (Brey, 2012a)

We contend that what Brey describes as the R&D stage is equivalent to what Moor identifies as the "pre-revolutionary" stage. During this R&D/pre-revolutionary stage, an anticipatory ethical analysis has the highest impact from a social and social justice perspective because it is at this stage that projections may be made for potential use cases for the technology. Brey agrees with this assessment:

> Since at the R&D and introduction stages many devices, usage patterns and social consequences are not yet present, ethical assessment turns speculative, as it focuses on particular

> R&D activities and techniques and then projects possible devices and usage patterns which are then assessed ethically. Such assessments may then be used to make ethical recommendations for R&D practices themselves, so as to increase the likelihood that these practices yield morally desirable devices and uses. Or they may be used for policy. (Brey, 2012a)

Thus, as technology develops, it moves from a research and development stage to a product introduction stage, to a market saturation stage, described as a power stage. At each of these stages of technological development, ethical and social problems may emerge as technology moves from stage to stage. According to Brey, as technology develops, there are a series of stages through which it passes:

1. The R & D and stage
2. The introduction stage
3. The permeation stage
4. The power stage

As we have already outlined, at the R&D stage, it is essential to understand the ethical impacts of the technology and the development of the technology in subsequent stages. According to Brey,

> In the introduction stage, few implementations of the technology exist, and the technology is still largely something that exists in labs and drawing boards. (Brey, 2012a)

At the introduction stage of technological development, the artifacts of the emerging technology begin to appear. Artifacts can be loosely defined as secondary and tertiary technologies that branch from and are related to the primary technological paradigm. These, as Verbeek and Vermaas state, are "clear-cut manifestations of technology" and are not, according to Hilpinen (2004), "unintended by-products of making (e.g., sawdust) or of experiments (e.g., false positives in medical diagnostic tests)" (Verbeek & Vermaas, 2009). It is currently unresolved as to whether ethical analysis can be extended to artifacts or whether they in themselves can be moral agents and is beyond the scope of this manuscript (Peterson & Spahn, 2010; Verbeek, 2009). During the initial stage, artifacts may arise as prototypes, and then improvements on prototypes emerge. The third stage is the permeation stage and is described as follows,

> In the permeation stage, the technological devices become more conventional and standardized. (Brey, 2012a)

At this stage of development, technologies and technological artifacts have matured far enough to enter the marketplace. Once technologies enter the marketplace, there is an additional stage of development, which is the power stage of technological development.

> In the power stage, the technology is firmly established. It is readily available, low- cost, and widely used, and many people in society are affected by it directly or indirectly. (Brey, 2012a)

In the power stage of technological development, technologies and artifacts are available to members of society at affordable prices. CRISPR has rapidly accelerated through Brey's stages 1 and 2 and is entering stages 3 and 4. This rapid movement into the later stages of Brey's paradigm has occurred with extraordinarily little analysis concerning anticipatory ethics. CRISPR has been accepted within the R&D community and has helped as a tool to make tremendous strides in accelerating our understanding of human genetics and disease. It is partially because of its stated promise to help ameliorate human disease that CRISPR has kept a critical analysis (speculative or otherwise) at bay, with some authors suggesting that critically analyzing these types of technologies for potential adverse effects as deleterious:

> [d]evelopments in synthetic biology have raised similar concerns, yet their expected perils and promises have so far failed to materialize as originally predicted. (Ben Ouagrham-Gormley & Vogel, 2016)

Because of the nature of CRISPR at the time of writing this manuscript, some aspects of the technology and its artifacts have progressed to stages 3 and 4, whereas others are still in 1 and 2. Based on Brey's assertion that the R&D stage (stage 1) is a speculative stage, where applications of the technology along with the artifacts are still being worked out, it is entirely appropriate to develop a critical yet speculative, anticipatory ethical analysis of the use of CRISPR in humans. Although "perils and promises have so far failed to materialize," it does not preclude them from materializing as artifacts and other prototypes spin out during the maturation of CRISPR through Brey's four stages. We would assert that even if a technology achieves Brey's fourth stage, these stages should not be viewed as static mileposts. Instead, Brey's stages should be seen as dynamic constructs, and even if a technology achieves the power stage, our anticipatory ethical analysis should not stop and may play a role in an ethical anticipatory governance policy model to guide the ongoing use of the technology in this stage and into the future.

The work of Deborah Johnson is vital in laying philosophical foundations for the development and expansion of anticipatory ethics. Indeed, we contend that Johnson's insights into the guiding principles of an anticipatory ethical analysis with her manuscript, "The role of ethics in science and engineering," represent one of the most succinct and foundationally important early frameworks for this analysis. Instead of looking at anticipatory ethics as a wrapper placed around developing technologies, Johnson suggests anticipatory ethical analyses as a parallel and integrated process within Moor's "technological paradigm." She defines it as such:

> Anticipatory ethics is a new approach that integrates ethics into technological development. (Johnson, 2010)

To emphasize the proactive, integrative, and cross-disciplinary (as opposed to the historically reactive and uni-disciplinary) nature of practicing anticipatory ethics, Johnson references work by Paul Verbeek (2010) to assert:

> Anticipatory ethics is ethical analysis aimed at influencing the development of new technologies. Advocates argue for ethics to accompany technological development rather than just reacting to it. (Johnson, 2010)

Johnson also states:

> Although a challenge, anticipatory ethics is the best opportunity for ethical perspectives to influence science and technology. (Johnson, 2010)

With these statements, Johnson reverses the order of the focus of ethical analysis. Instead of waiting for ethical problems to emerge after a technology has developed and after it has been implemented, an *ethical analysis should accompany and anticipate the development of technology*. Perhaps, most importantly, an anticipatory ethical analysis should be thought of as *helping to influence the development of science and technology*. The idea that ethical analysis can be used to help guardrail science is not new; what is, however, is the notion that we should develop anticipatory frameworks that operate in parallel, or slightly front-run technological paradigms that are in Moor's "pre-revolutionary" or Brey's "R&D" stage. Although some may see this as needlessly speculative, we argue that this type of anticipatory approach is needed to keep pace with the rapid and exponential growth of technologies in general and biomedical technologies in specific, like CRISPR that stand to rapidly progress to both the "power" and "revolutionary" stages. This is important to guide the conduct and scope of scientific endeavors towards a common social good. Johnson's insight here is apropos:

> Moral notions, principles and practices influence the conduct as well as the outcomes of science. The requirement for informed consent of research subjects is perhaps the best example of a moral principle that shapes the conduct of research. Conversely, science and engineering have powerful effects on morality; computers have challenged ideas about what it means to be human; reproductive technologies have made parenthood seem like a form of consumption; software has changed the rights associated with property; and pharmaceuticals have solidified ideas about what it means to be normal. (Johnson, 2010)

What is essential from these passages about anticipatory ethics are the orientation of the starting point and the trajectory of the ethical analysis. As stated, anticipatory ethics stands in opposition to a reactionary approach to ethics. Within the context of technological development, this ethical framework represents a significant move away from reacting to technology or technological artifacts after something has gone wrong, towards looking at potential problems with the development of technology and artifacts at the R & D stage of development, before and as they are being designed and brought into existence.

The combination of the viewpoints of Moor, Brey, and Johnson develop a robust framework for anticipatory analysis of a technological paradigm, technology, or artifact, but what about the subjective experiences of those developing and using that technology? An anticipatory ethical analysis can be brought to bear in this case. Let us simply refer to this as an anticipatory "subject-oriented approach." An anticipatory subject-oriented approach to ethical problems begins with how artifacts influence lived experiences. For Brey:

> Ultimately, ethical assessment of emerging technologies concerns the question of what is good and bad about the device and processes that they may bring forth, and what is right and wrong about ways in which they may be used. (Brey, 2012a)

The ideas of "good," "bad," "right," and "wrong" can be viewed through a multitude of ethical lenses. These lenses can be easily focused on CRISPR, and as an emerging technology, we need to ask what is good and bad about the technology and the right and wrong ways it can be used. What is clear is that we must pivot away from ethical viewpoints couched in a classical historical/transcendental approach to the philosophy of technology.

> The classical philosophers of technology occupied themselves more with the historical and transcendental conditions that made modern technology possible than with the real changes accompanying the development of technological culture. (Achterhuis, 2001)

This "transcendental conditions" approach to the philosophy of technology and ethical analysis undermines an ethical framework that performs analysis and exploration where technology finds expression through the subjects that interact with the technology. While an anticipatory ethical analysis of CRISPR could begin by discussing what is good and bad about the technology and what are the right and wrong ways it can be used in a historical/transcendental philosophical context, we posit that with rapidly advancing technologies and aligned with previous work from others, that CRISPR requires an anticipatory subject-oriented approach.

Anticipatory ethics is indeed concerned with the need to recognize the importance of the insights of traditional philosophy of technology and the need to recognize the importance of empirical understanding of social and ethical issues related to the development of the lived experience of the users of the technology. As this technology and the associated technological artifacts develop, they affect existence at individual and societal levels. This is true of rapidly evolving revolutionary technologies like CRISPR and will be the case for the iterations of CRISPR that continue to develop. As stated earlier, it is uncontroversial to posit that CRISPR continues to be applied to an ever-larger group of areas within scientific research and development. More sophisticated iterations of CRISPR and other gene-editing technologies will also continue to occur.

It is only through the anticipatory study of technology and empirical case studies that emphasize the lived experiences of others that we can observe the

> …ways in which technology has transformed – and continues to transform – social networks and forms of life, human needs and possibilities, and the experience of our bodies and nature. (Achterhuis, 2001)

So, by examining how technology and technological artifacts (such as CRISPR) develop within the lived experience and how technology and technological artifacts are used in a social context, we can identify the ethical and social problems with technological paradigms, technology, and artifacts. It is upon this basis technology transforms individual and social existence. We can also begin to see how three distinct orientations are developed for anticipatory ethical analysis.

References

Achterhuis, H. (2001). *American philosophy of technology: The empirical turn*. Indiana University.

Ben Ouagrham-Gormley, S., & Vogel, K. M. (2016). Gene drives: The good, the bad, and the hype. Retrieved from https://thebulletin.org/2016/10/gene-drives-the-good-the-bad-and-the-hype/

Brey, P. A. (2012a). Anticipatory ethics for emerging technologies. *NanoEthics, 6*(1), 1–13. https://doi.org/10.1007/s11569-012-0141-7

Brey, P. (2012b). Anticipating ethical issues in emerging IT. *Ethics and Information Technology, 14*, 305–317.

Demirci, S., Leonard, A., Haro-Mora, J. J., Uchida, N., & Tisdale, J. F. (2019). CRISPR/Cas9 for sickle cell disease: Applications, future possibilities, and challenges. *Advances in Experimental Medicine and Biology Cell Biology and Translational Medicine, 5*, 37–52. https://doi.org/10.1007/5584_2018_331

Hester, K., Mullins, M., Murphy, F., & Tofail, S. A. (2015). Anticipatory ethics and governance (AEG): Towards a future care orientation around nanotechnology. *NanoEthics, 9*(2), 123–136. https://doi.org/10.1007/s11569-015-0229-y

Hilpinen, R. (2004). Artifact. In E. N. Zalta (Ed.), *The Stanford encyclopedia of philosophy* (Fall 2004-Edition). http://plato.stanford.edu/archives/fall2004/entries/artifact

Johnson, D. G. (2010). The role of ethics in science and engineering. *Trends in Biotechnology, 28*(12), 589–590. https://doi.org/10.1016/j.tibtech.2010.08.003

Nanayakkara, P., Diakopoulos, N., & Hullman, J. (2020). Anticipatory ethics and the role of uncertainty. arXiv:2011.13170.

Nestor, M. W., & Wilson, R. L. (2019). An anticipatory ethical analysis of robotic assisted surgery. *Business and Professional Ethics Journal, 38*(1), 17–42. https://doi.org/10.5840/bpej201812675

Nestor, M. W., & Wilson, R. L. (2020). Beyond Mendelian genetics: Anticipatory biomedical ethics and policy implications for the use of CRISPR together with gene drive in humans. *Journal of Bioethical Inquiry, 17*(1), 133–144. https://doi.org/10.1007/s11673-019-09957-7

Nestor, M. W., Artimovich, E., & Wilson, R. L. (2014). The ethics of gene editing technologies in human stem cells. *Ethics in Biology, Engineering and Medicine: An International Journal, 5*(4), 323–338. https://doi.org/10.1615/ethicsbiologyengmed.2015013732

Nikolova, B. (2013). The rise and promise of participatory foresight. *European Journal of Futures Research, 2*(1). https://doi.org/10.1007/s40309-013-0033-2

Peterson, M., & Spahn, A. (2010). Can technological artefacts be moral agents? *Science and Engineering Ethics, 17*(3), 411–424. https://doi.org/10.1007/s11948-010-9241-3

Pimple, K. D. (2014). *Emerging pervasive information and communication technologies (PICT): Ethical challenges, opportunities and safeguards*. Springer Netherlands.

Plater, R. (2020, July 06). First person treated for sickle cell disease with CRISPR doing well. Retrieved from https://www.healthline.com/health-news/first-person-treated-for-sickle-cell-disease-with-crispr-is-doing-well

Regalado, A. (2021, October 20). Exclusive: Chinese scientists are creating CRISPR babies. MIT Technology Review. Retrieved April 18, 2022, from https://www.technologyreview.com/2018/11/25/138962/exclusive-chinesescientists-are-creating-crispr-babies/

Verbeek, P. P. (2005). *What things do: Philosophical reflections on technology, agency, and design*. Penn State University Press.

Verbeek, P. (2009). The moral relevance of technological artifacts. In *The international library of ethics, law and technology evaluating new technologies* (pp. 63–77). https://doi.org/10.1007/978-90-481-2229-5_6

Verbeek, P. (2010). Accompanying technology. *Techné: Research in Philosophy and Technology, 14*(1), 49–54. https://doi.org/10.5840/techne20101417

Verbeek, P. P. (2011). *Moralizing technology: Understanding and deigning the morality of things*. The University of Chicago Press.

Verbeek, P., & Vermaas, P. E. (2009). Technological artifacts. In *A companion to the philosophy of technology* (pp. 165–171). https://doi.org/10.1002/9781444310795.ch28

Wang, H., La Russa, M., & Qi, L. S. (2016). CRISPR/Cas9 in genome editing and beyond. *Annual Review of Biochemistry, 85*, 227–264. https://doi.org/10.1146/annurev-biochem-060815-014607. Epub 2016 Apr 25. PMID: 27145843.

Chapter 3
CRISPR, Phenomenology, and Lived Experience

3.1 Introduction

This chapter will describe how CRISPR will create alterations in medical research and medical practice and anticipate some ethical issues that may arise for scientific/medical research, clinicians, and patients due to the commonplace adoption of CRISPR gene-editing technology. The methodology developed will follow a specific path. The inspiration for this analysis will be based upon the phenomenological philosophy of Edmund Husserl. This is due to the central role that lived experience plays in Husserl's philosophy. A discussion of postphenomenology will follow this. An account of the post phenomenological philosophy of Don Ihde will be introduced, and both of these approaches will be conjoined with an account of anticipatory ethics. Distinctions drawn from these sources will be combined with our interpretation of how these key concepts from phenomenology and postphenomenology can be connected to basic theories of philosophical ethics. A stakeholder approach to ethical analysis will be introduced to follow a principle-based method that focuses on applying ethical principles involving three groups of stakeholders—scientific researchers, clinicians, and patients. At the center of our analysis is the lived experience of each of these stakeholders. While many theoretical viewpoints will be introduced, it is essential to remember that this discussion aims not to make a wide-ranging scholarly contribution in each area but to employ these ideas and concepts as part of the foundation we believe is necessary for understanding ethical issues with CRISPR.

The inspiration for this analysis is the phenomenological philosophy of Edmund Husserl. We start with phenomenology due to Husserl's emphasis upon *lived experience*. The goal of the analysis presented here is to use distinctions from Husserl's philosophy to clarify both technical issues related to how stakeholders understand CRISPR and ethical issues with the use of CRISPR. Here, we hope to use the combination of an anticipatory ethical analysis combined with an applied phenomenological approach to underscore the idea that uses of CRISPR need to consider stakeholder perspectives of both present and future persons from the context of their

M. W. Nestor, R. L. Wilson, *Anticipatory Ethics and The Use of CRISPR in Humans*, https://doi.org/10.1007/978-3-030-98368-0_3

lived experiences. In essence, our analysis here is applied phenomenology to highlight the anticipatory ethical issues surrounding the future use of CRISPR in humans.

Husserl's specific distinctions will guide this analysis, which as was stated, amounts to an exercise in applied phenomenology. Consequently, there will be no effort made to develop Husserl scholarship and illustrate in detail all the technical distinctions introduced in Husserl's phenomenology. The distinctions that will be employed will be described only to the degree that is needed to carry out the development of a foundation for the anticipatory ethical analysis of CRISPR. This will require that we *interpret* the presentation of Husserl's ideas in his published works. These ideas are well suited to help understand stakeholder perspectives and ethical issues with emerging and disruptive technologies like CRISPR that are currently undergoing exponential growth in what has been called the "4th industrial revolution" (Schwab, 2017). This is in contrast with other styles of analysis in that Husserl's approach enables each of the stakeholders to understand one another through an empathetic interpretation of their bodily and cognitive concerns (Husserl, 1990).

Our assertion here that an important principle that underlies and unifies most of the different stakeholder perspectives in this anticipatory analysis is the premise that CRISPR is an increasingly successful gene-editing technique that *will extend into the future*. It is essential to recognize that judgments about the success of CRISPR now will be combined with the projection of *imaginary views* of how CRISPR will impact and affect the future and influence the direction of scientific research and the world of future health care (Drummond, 2015). Our analysis does not offer a specific prediction about the future state of the world and CRISPR, as much as issue a challenge to stakeholders to engage in imaginative anticipatory projections from their perspectives to engage the future concerning the use of CRISPR and then thinking about what future world they want to build by acting in the present. This imaginative anticipatory projection must require constructing models of future lived experience based on a heuristic approximation of a future with CRISPR. An overview of some aspects of Husserl's phenomenology will be helpful here.

There are a variety of entry points into Husserl's phenomenology. There will be four main concepts from Husserl's work that will provide the foundations for the present analysis (Fig. 3.1). These concepts include the following:

- The life world
- The natural attitude
- Reflection
- Intersubjectivity

To help understand how Husserl's ideas apply to CRISPR, we will employ our three anticipatory stakeholder perspectives to give a preliminary understanding of how phenomenological ideas and concepts can be applied to the scientific development of CRISPR. We assume patients suffering from a genetic condition have a specific lived experience that involves searching for a treatment or cure for the condition. We presume one significant aspect of the clinician's lived experience is to eventually use CRISPR to alleviate the suffering of patients by correcting genetic conditions that contribute to that suffering. We assume that the lived experience of

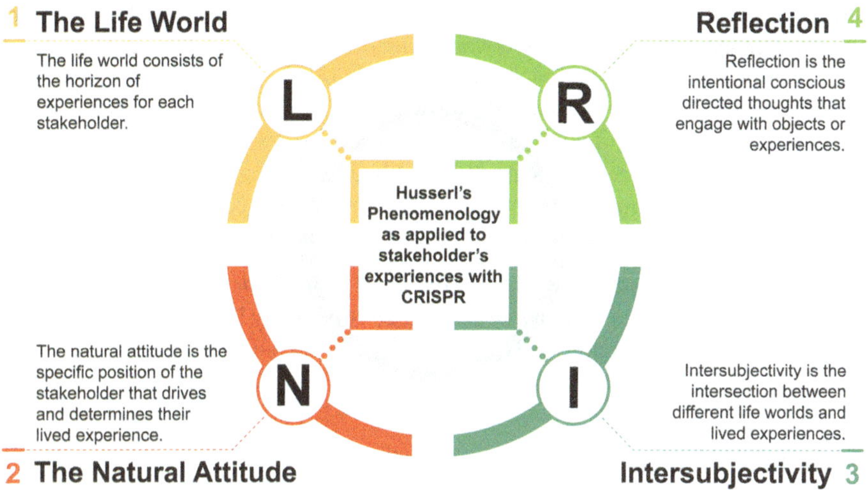

Fig. 3.1 Interrelated fundamental main concepts of Husserl's phenomenology used as foundations for the current anticipatory analysis

researchers is concerned with conducting scientific research to develop CRISPR as a gene-editing technology, but also to provide the tools by which the clinicians can potentially alleviate the problems experienced by the patients due to their genetic condition.

3.2 Life World (*Lebenswelt*)

We will start with a fundamental concept for Husserl's phenomenology—the *Lebenswelt* or "life-world." For each stakeholder, the surrounding life-world is the ever-present horizon of experiences, and this horizon is the basis for all assumptions, the backdrop for all "positing." To posit is to live with the assumption and certainty that the world exists as experienced (Husserl, 1970; Moran et al., 2012). For a patient to posit is for the patient to take it for granted that they live in a world where they suffer from a genetic condition. We all live within a world of consciousness that forms a background to our experiences, thoughts, and feelings. Our life-world is the subjective pre-given world within which we find ourselves and a world of immediately perceived experiences about a world that is always already there. Husserl refers to this pre-given conscious dynamic horizon of objects and time that make up the life-world as the ground state for intersubjective experiences. It is these intersubjective experiences and perceptions that humans share as a common world, in which some level of mutually endorsed objective truths can be discerned.

There are some limits to the concept of life-world experiences, however. Life-world perceptions within a common world are typically experienced in a straightforward unreflective fashion. In each of our lived experiences, we simply live through existence in a common world with others as objects in time, in a "grand theater of objects" (Stanghellini & Mancini, 2017). This means that, although setting the ground for a common world, life-world perception and experience is still a personal consciousness of a world—albeit subjected to the historical, cultural, and social frameworks of meaning superimposed by the common world. For the stakeholders that CRISPR impacts (in our case here, we are focused on researchers, clinicians, and patients), we can make two general assumptions:

1. They all share a common world that has as its ground a *lebenswelt* experience.
2. They perform their tasks in a *lebenswelt* that is typically lived through and experienced relatively unreflectively.

3.3 The Natural Attitude

Husserl's concept of the "natural attitude" is directly related to the notion of the life-world. In the natural attitude, each of us takes for granted both the existence of the world in which the objects of our experiences are thought to exist and the validity of our judgments about those objects. A major characteristic of the natural attitude is what Husserl calls its "general thesis."

> The general thesis is the belief that the world before me is a factual existent world, and that this world represents itself as a factually existent actuality. (Drummond, 2010)

Thus, the natural attitude refers to our natural way of being in the world,

> By my seeing, touching, hearing, and so forth, … corporeal physical things … are *simply there for me,* 'on hand' … whether or not I am particularly heedful of them. (Husserl, 1990)

We exist in a common world that is taken for granted because we live in it without reflecting upon its status. We just assume the external world exists and that the things within the world exist as we perceive them to exist.

> First, the natural attitude is not a self-evident fact (as we see with many philosophers); rather it is a stance, an approach to how we experience things in the world. In his description of it Husserl points out other attitudes each of which takes up its own types of objects, each with its own essential rules. (Jacobs, 2021)

For each of our exemplary stakeholders, lived experience is characterized by a specific experience in the natural attitude. For patients, there is lived experience influenced by a genetic disorder. For clinicians, the idea of alleviating the patient's condition is part of their lived experience. For scientific researchers, the concept of perfecting gene editing is central to their lived experience. Each attitude, then, is a particular position that affects how each stakeholder sees things—and what they see (Husserl, 1990). Thus, based on the natural attitude of each stakeholder within their life-world, the lived experience of each of our exemplary stakeholders is potentially non-overlapping concerning CRISPR. However, this is not the case for everyone within each group of stakeholders. These stakeholders have an attitude that is altered by the possibilities of developing CRISPR amongst other gene-editing technologies. The patient, through gene editing, can be treated or cured of a genetic condition. The clinician can alleviate the suffering of a patient. The researcher can alter the world and the world of the future through the continually evolving development of CRISPR. These all represent different life-worlds, but due to the development of gene-editing techniques such as CRISPR, they share a common intersubjective life-world. Thus, our earlier premise #2 is not valid for *all* stakeholders, only some. Stakeholders in an anticipatory ethical analysis must be assumed to have a life-world perception of CRISPR based on a natural attitude that drives each stakeholders' lived experiences. Some stakeholders move through their life-world in an unreflective way, and some within these groups reflect on their phenomenological experience of their intersubjective world, which Husserl defines precisely.

3.4 Intersubjectivity

Intersubjectivity is an essential concept for Husserl. The world that any individual perceives is not just a world within that person's mind but a world that others can access through their lived experience. (Gallagher, 2012). The idea that each stakeholder can share in an intersubjective world potentially shared with each other stakeholder and reflect on the lived experience of the other can be called a form of "transcendental intersubjectivity":

> Transcendental intersubjectivity is the absolute and only self-sufficient ontological foundation out of which everything objective (the totality of objective real entities, but also every objective ideal world) draws its sense and validity. (Husserl, 1980)

One way to think about intersubjectivity in the context of the life-world, natural attitudes, and the lived experience is to think about how the different worlds of stakeholders intersect due to the intersubjective nature of the meaning of CRISPR is constituted for each of them. Each of the individual stakeholders has a unique view of CRISPR: as the object of scientific research, as a technology that can be used to intervene and correct a person's genetic condition, and as a means to alleviate the suffering associated with a genetic condition. Even though the object in their lived experience is different for each, these perspectives can all be combined by each of the stakeholders, and in the aggregate, all refer to the same gene-editing technique. In this way, the same intersubjective experience can result in a vastly different constitution of each stakeholder's meaning with respect to lived experiences. In contrast, each stakeholder can agree that their own lived experiences can be brought to bear on a common understanding of what CRISPR is and what it can do. What seems evident is that all stakeholders can constitute the meaning of gene editing intersubjectively. The lived experience of an individual suffering from a genetic condition is uniquely experienced by that individual. At the same time, intersubjectivity is capable of being understood as part of the lived experience of the clinicians and researchers when assessed by them (Fig. 3.2).

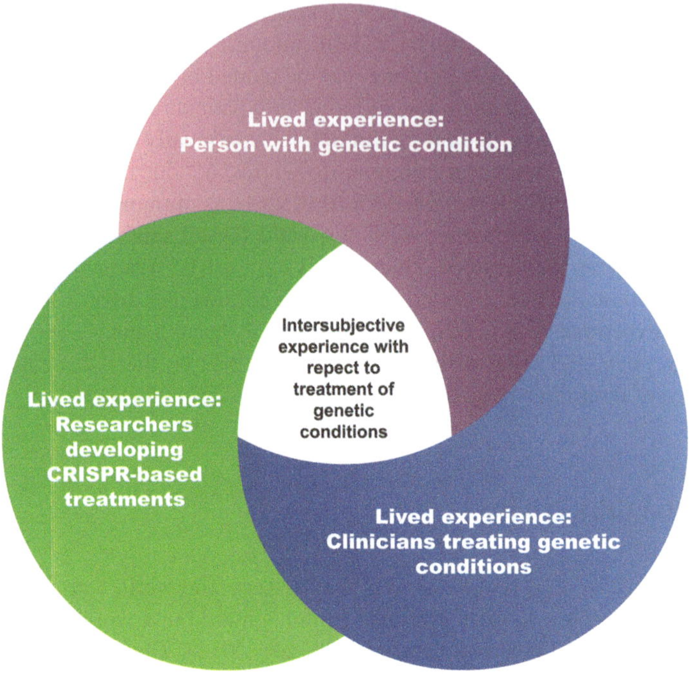

Fig. 3.2 Intersubjective experience requires the intersection between the lived experiences of all stakeholders, including patients, clinicians, and researchers. Each stakeholder has their own inter-subjective experience of CRISPR technology:

> The objectivity and transcendence of the world – its otherness, the fact that it does not reduce to just a set of ideas in my mind – is constituted by this intersubjective dimension. The world is not just my world since it is experienced by others in ways that do not reduce to my own experience of it. (Gallagher, 2012)

The point of presenting this phenomenological background is to emphasize the nature of the lived experience of the stakeholders who are at the center of CRISPR gene editing technologies. There are those who engage in the scientific research related to CRISPR, those who will employ CRISPR for medical interventions, and those who will have their genetic conditions corrected by CRISPR. Each of our stakeholders exists in an intersubjective shared world that is characterized in the natural attitude by causal relations. Genetic conditions result in behavioral and physical phenotypes. Clinical interventions that use CRISPR can potentially correct genetic conditions and cause the alleviation of suffering. In the natural attitude, an individual's lived experience and life-world are characterized by finding oneself in an environment where gene editing objects are taken to exist and with which they interact. When we think about the gene-editing objects we find in the physical world, we think about them in terms of cause-and-effect relationships. In this way, stakeholders with different lived experiences can converge on what CRISPR is and what it can do. More specifically, it can reflect on the cause-and-effect relationship between genetic conditions, physical symptoms, and tools to alleviate those symptoms. Scientific research can perfect the process of genetic interventions. In the intersubjectively shared world, these stakeholders can engage in imaginative reflection, which we will now discuss. These reflections also make up the intersubjective experience of stakeholders. A patient who suffers from a genetic condition can imagine a world free from symptoms. A clinician can imagine a world where CRISPR has migrated to the clinic and can now be used as a tool to help patients. A researcher can imagine a gene-editing technique that has been perfected as a research tool for discovery and can also be used in a clinical application.

3.5 Phenomenology and Anticipatory Stakeholders

The phenomenological framework that we have described sets the stage for understanding stakeholders' perspectives in an anticipatory ethical analysis of CRISPR. The application of phenomenology to issues related to health care is not new and has been employed by numerous authors (Carel, 2010; Mapp, 2008; Toombs, 2001; Zaner, 1981). Phenomenological bioethics (as this application of phenomenology to medicine and health care is called) is focused on an ethical understanding of issues involving patients, including:

> ...carrying out investigations of the lived experiences of the persons involved in morally problematic situations in order to see how the main principles of [ethics and] bioethics will apply. (Svenaeus, 2018).

To deepen our understanding of the interdependence of the lived experience of each of our groups of stakeholders, we need to introduce the concept of reflection in Husserl's phenomenology. The phenomenological experiences comprising the life-world, the natural attitude, and intersubjectivity all allow each stakeholder to engage in a reflective action towards their own lived experience concerning CRISPR. Within a Husserlian framework, this reflective action is an intentional act focused on the intention of CRISPR technology itself. In Husserl's formulation, defined as the "noema" or "noematic" aspect of the experience of CRISPR. Husserl's extensive analysis of the role of the noema, which goes substantially beyond simply calling the noema an 'object', is beyond the scope of our analysis. The reflection upon intention is experienced differently by each stakeholder based on their life-worlds and how their life-worlds intersect intersubjectively with the life-worlds of other stakeholders.

All of our stakeholders occupy the same intersubjective space in the shared overarching life-world. Still, they also have a unique phenomenological perspective driven by immediate lived experiences within their life-world. At this point, it is important to recognize that it would be absurd to think that all these stakeholders naively blunder through an accepted life-world with little or no thought or reflection. Reflection is then a natural flow from the intersection and intersubjectively shared life-world. This world is one in which each stakeholder reflects upon their lived experience-and we assume this occurs for most stakeholders. We go so far as to say that sophisticated reflection (reflection that is deeper than a reflection upon the immediate experience of the life-world), and which extends beyond the shared life-world, occurs for most stakeholders in an intersubjective fashion.

How would this look in practical terms? Based on Husserlian phenomenology, we can make some assumptions about the lived experience of a set of stakeholders using a tightly bounded scenario developed for discussion. The scenario is a parent who has discovered during pregnancy that the baby has an inherited genetic condition that can be potentially alleviated through CRISPR. This parent has decided to consider individualized gene therapy treatment by subjecting the fetus to a CRISPR construct delivered through the use of an adenovirus. For the sake of discussion, we will assume that the adenovirus construct can be delivered in a targeted fashion to the organ of interest implicated in the genetic condition.

A high-level reflective assessment of the possible immediate lived experience of each stakeholder can be performed. We will walk through this framework from the perspective of the parent of a child with a genetic condition and the child. This assessment requires one last Husserlian concept, which we will introduce now. This is the concept of the *intentionality of consciousness*. The intentionality of consciousness suggests that reflection from the perspective of a stakeholder's lived experience requires a retrospective and prospective analysis that entails reflection on the past and *projection into the future*. This is because Husserlian intentionality posits that:

> our experience is directed toward—represents or 'intends'—things only through particular concepts, thoughts, ideas, images, etc. These make up the meaning or content of a given experience, and are distinct from the things they present or mean. (Smith, 2013)

For the parent of a child who has an inherited genetic condition, there will be a detailed reflection upon the genetic condition (in this case, this would be Husserl's *noema* or the "object" at which acts of consciousness are directed). There are acts of consciousness (intentionality of consciousness) related to addressing and living with the condition. The child (if older and able to consciously reflect) may also have a similar reflective experience to that of the parent, with the added layer of the lived experience of the genetic condition as a part of their life-world. Bringing in the reflective lived experiences of clinicians and researchers, we can broaden the phenomenological framework from a stakeholder perspective.

The *noema* for the clinician is both the genetic condition and the tool used potentially to treat the condition, in this case, CRISPR. Clinicians may then have intentional acts of consciousness directed toward how to address the condition using clinically available treatments and a clinical understanding of the genetics and phenotypic readouts of the condition itself. The researcher in this framework may have a more scoped *noema*. In the researcher's case, a targeted and effective CRISPR construct that can be used in a clinical setting would be an intentional act of consciousness and detailed reflection. Other examples can be used within the reflective scope of each of these stakeholders. Still, these illustrate how the life-world intersects with the intentionality of consciousness and directed reflection on a *noema* to guide each stakeholder's lived experience and phenomenology in a future-oriented direction. The patient sees a future without a genetic condition. A clinician sees CRISPR-driven interventions to correct a genetic condition while researchers are directed towards perfecting gene-editing techniques.

The phenomenological experience of each stakeholder can also be exposed to the intersubjectivity of the lived experiences of the other stakeholders, and there can also be an imaginative reflection through intersubjectivity upon the experience of the other. The intersubjective experiences of stakeholders can provide a framework for them to formulate beliefs and desires based upon acts of consciousness related to gene editing in general and CRISPR specifically. Once stakeholders formulate beliefs and desires based upon reflection and imaginative reflection based on intersubjective experiences, they can then begin to project these beliefs and desires onto the perspectives of other stakeholders.

Each stakeholder's intersubjective experience and imaginative reflection require the same retrospective/prospective process discussed earlier. The projective aspect of the process can have a varying set of time horizons. We can generally distinguish them as short-term, intermediate-term, and long-term. Although the idea of an intermediate-term time horizon is admittedly fuzzy, we can, for the sake of discussion, think of short-term as, say 0–5 years; intermediate-term as 5–10 years, and long-term as 10–25 years into the future. These are arbitrary distinctions, but they will serve the purpose of bounding the prospective process of each stakeholder.

The following example proceeds from each stakeholder's imagined reflection based on intersubjective experience, resulting in a set of beliefs about CRISPR that can be projected onto the perspectives of other stakeholders. For each of the stakeholders, once the *noema* is identified as the genetic mutation, acts of consciousness begin to emerge, which can lead to beliefs from each stakeholder's perspective about CRISPR. Depicted below is a *non-exhaustive* list of acts of intentional consciousness and potential beliefs for each of our selected stakeholders in the short-term horizon.

Patient	Awareness of the genetic condition.
	In a trial for pure research.
	In a trial for a therapeutic intervention.
	Hope for a somatic intervention correcting a genetic condition.
Clinician	Aware of clinical tools available.
	Alleviate the suffering of the patient.
	The normalcy of function for the patient.
	Enhancement of health for the patient.
Researcher	Development of genetic editing technique.
	Gene editing technique deployed.
	Further developments based on current deployments.
	Enhancement of gene-editing techniques.

Thus, the patient (and parent in this case) is made aware, by health care professionals, of a genetic condition caused by a mutation in a gene or set of genes in the patient. Without a genetic engineering intervention, the patient will suffer and potentially die from the genetic mutation. This sets off a cascade of reflections (retrospective, present, and prospective) for both the parent and child that results in actions taken by the parent to enroll the child in either a trial therapy or clinically approved therapy for the condition. In this example, the therapy proposed is a CRISPR-driven therapy that both the parent and child hope will result in a cure or significantly reduced symptoms. For the clinician, the clinical application of the correct gene-editing tool can affect their patient's health status. With a CRISPR gene-editing intervention, patients have the potential to be "cured" of the genetic condition. This will alleviate suffering and enhance through improved health the life of the child. For the researcher, developing a perfect gene-editing tool with no off-target effects that can be directly targeted to organs or cells of interest is ideal, an ideal that, if achieved will affect both the patient's and clinician's lived experience.

The intentionality of consciousness combined with the intersubjective phenomenological lived experiences of each of the stakeholders, in this case, is buttressed by a belief and desire that a perfect gene-editing technology that can identify and target genetic mutations, remove, and replace them, can be developed and can be done without side effects and unintended consequences. This is a belief driven by each stakeholder's lived experience. Each stakeholder has a unique perspective from which they approach CRISPR, and each is directed towards and thinks about gene editing in general from their unique perspective. Each stakeholder is preoccupied with their individual concerns and solutions based upon their unique lived experience in the short-term time horizon. At the same time, each stakeholder refers to the same *noema*. At this point in the analysis, it would not be controversial to say that there would be scant ethical concerns about moving forward with the CRISPR-driven adenovirus approach to help the child.

3.6 Postphenomenology

The phenomenological framework for understanding each stakeholder's perspective in an anticipatory analysis underscores the fundamental presence of each stakeholder's lived experience and how those experiences intersect with the intersubjective experiences of the other stakeholders to establish beliefs about technology or technological artifacts. This complex interaction requires reflection of both a retrospective and prospective nature. Although these experiences can be focused on the same noema, Husserlian phenomenology sets a relationship between the stakeholders and the noema that is distant and treats the noema (in our case, this could be the genetic condition or CRISPR) as "the other," effectively alienating the lived experience of the stakeholder from the technology itself. Some aspects of a postphenomenological framework can be applied to the relationship between the stakeholder and the technology to understand why there is a complex interplay between the technology itself and the prospective aspect of reflection for each of the stakeholders.

To understand how this may be the case, we will first briefly conceptualize postphenomenology. Then, we will use concepts from Don Ihde to clarify distinctions that are central to our analysis of CRISPR without entering into a deeper philosophical analysis of these concepts. Ihde has developed postphenomenology to operationalize phenomenology in a practical context, emphasizing the technology and the human experience of that technology. In other words, the *noema* is no longer a technological artifact that each stakeholder reflects on as a static "other" but as a dynamic aspect of the lived experience that can affect the relationship between humans and the technological artifacts they encounter. The study of cases using this methodology focuses on how humans interact with technology and technological artifacts and how this, in turn, influences how people interact with one another:

> Postphenomenologists study technology in terms of the relations between human beings and technological artifacts, focusing on the various ways in which technologies help to shape and mediate between relations between human beings, technologies, and the world. They do not approach technologies as merely functional and instrumental objects, but as mediators of human experiences and practices. (Irwin, 2017)

Technological artifacts influence deliberations, choices, and the performance of actions and experiences in the world. Postphenomenology can be employed to study artifacts and see how these artifacts change day-to-day life-world experiences. These artifacts shape the intentions of both designers and users and those upon whom they are employed. What is critical in postphenomenology is the study of the relationships that develop between technologies, the users of these technologies, and those upon whom technological artifacts are used. Technologies and technological artifacts influence the lived experience of those who employ them and upon the lived experience of those upon whom the artifacts are used. An essential insight of postphenomenology is the recognition that:

> Technological transformations…are non-neutral because these technologies are not neutral … over time we become conditioned by our technologies in non-neutral ways. (Irwin, 2017)

CRISPR is not a neutral technology when employed to alter genes and gene expression. Technologies and technological artifacts may be used as tools, but they are tools that shape our experience of the world. For example, the development of computer technology has fundamentally changed how information is accessed, how this information influences stakeholders, and how we communicate with one another. The postphenomenological perspective can be employed to recognize that, in practical terms, medical technology (like computer technology) influences medical practices. However, new medical technologies also mediate the relationships between medical practices, staff, patients, and researchers. The analysis goes deeper than this. According to Ihde, a phenomenology of technology needs to focus on the structural features of human-technology relationships. Ihde's analysis centers on how we are engaged in a bodily fashion with technologies (Ihde, 1990).

These are four ways that Ihde's postphenomenology expands upon phenomenology to discuss how technology and artifacts play a role of 'mediation' within lived experience. These distinctions broaden and build upon the noesis-noematic distinctions of Husserl. Ihde calls these categories:

- Embodiment relations
- Hermeneutic technics
- Alterity relations
- Background relations

A thorough discussion of each of these categories is beyond the scope of this manuscript. In the context of CRISPR and humans in our analysis, embodiment and background relations are essential aspects of Ihde's framework and can be used to analyze a stakeholder's intersubjective experiences. The relationship between technologies and human bodies can lead to the embodiment of technological artifacts and influence stakeholders' retrospective and prospective reflection.

In embodiment relations, technologies influence lived experience because they enter into a mediational unity with human lived experience, and this unity is directed at the world. We see the world through the glasses when we wear glasses, but we don't see the glasses themselves. The technology is absorbed into our lived experience. This "absorption" is defined as a quasi-transparent technology. Ihde describes this in the following way:

> I call this first set of existential technological relations with the world embodiment relations, because in this use context I take the technologies into my experiencing in a particular way by way of perceiving through technologies and through reflexive transformation my perceptual and body sense. (Ihde, 1990)

In embodiment relations, we take the technology as an artifact that, over time, is absorbed and fades into the lived experience of the stakeholder and becomes a lens through which the stakeholder perceives their lived experience and the intersubjective experiences of others. This becomes critically important in the case of CRISPR. As CRISPR is used to correct genetic conditions, it also alters the life world of each of these stakeholders and can take on an embodiment role for the stakeholder who is using CRISPR as a treatment. When that treatment is a correction for a genetic condition that can lead to a somatic phenotype, such as sickle cell disease, the embodiment is much like the example of glasses. Over time, the person using the glasses benefits from the treatment, which becomes transparent like the technology itself. This transparency allows the technology to carry over into other areas of life to enhance experience. When the use of CRISPR is a correction or enhancement for a genetic condition that affects the human mind, then the transparency of the treatment has not only the effect of the glasses, but quite literally alters the phenomenological experience of the lived world to such an extent that the idea of transparency itself must be contextualized. Transparency in the case of glasses or CRISPR for sickle cell presumes the stakeholder who receives the treatment can articulate a reality without each realizing the technology has been absorbed into the lived experience. When CRISPR is used to alter the mind of the stakeholder and becomes mentally embodied, then it stands to reason that there are two possible outcomes:

1. The stakeholder can infer a phenomenological reality without CRISPR by relying on the intersubjective experiences of stakeholders who have not been treated, or
2. A world without CRISPR is unimaginable for that stakeholder because genetic engineering has been absorbed into lived experience, preventing the stakeholder from articulating a reality in which CRISPR modifications do not exist.

If the mind of a stakeholder has been modified so that the modification is fully mentally embodied, altering the perception of the modified stakeholder, then #1 may be possible, but #2 is equally likely. This construct thus flows into the modified stakeholder's reflective analysis. But it does not stop there. Just as with someone wearing eyeglasses that are not perceived by the person seeing through them-both the clinician and researcher absorb CRISPR gene-editing technology into their lived experiences. Here the idea of using CRISPR as a gene-editing intervention is no longer explicitly noticed; it is simply employed. This has two implications.

First, suppose the stakeholder that has been mentally modified is using the intersubjective experiences of the other stakeholders to infer a different lived experience. In that case, this perspective may no longer be available to the modified stakeholder. Second, if all stakeholders are relating to CRISPR in a transparent way, then the prospective aspect of their reflective analysis may not consider the lived experience of future stakeholders. Postphenomenological claims about technology are posed from an embodied and situated perspective. In the case of CRISPR, the lived experience of each of the stakeholders is affected by the relationship the stakeholder has to the CRISPR gene-editing technology, so it is reasonable that reflection to assess the suitability of CRISPR use in future persons would be affected by the mental embodiment of CRISPR in the patient. Interpretations of current uses of CRISPR for mental enhancement would project present ideas about cognitive enhancement unto the lived experience of future persons. For the clinician and researcher, the embodiment relation is less clear. What would apply to these stakeholders is the concept of Ihde's background relations, these are:

> …technologies that are part of the environment we live in. Examples are a thermostat or refrigerator running in the background (Rosenberger and Verbeek 2015, p. 18). There is a move beyond transparency here, in that these technologies are a 'present absence', shaping our environment without requiring direct action. (Rosenberger & Verbeek, 2015)

It is reasonable to assume that CRISPR will manifest as a background relation for clinicians and researchers after enough use. Thus, in the case of these stakeholders, their intersubjective lived world may not allow for meaningful reflection of CRISPR in either a retrospective or prospective fashion. When genetic engineering has become part of the taken-for-granted background of current lived experience, what those who will exist in the future will think about our genetic modifications that will condition their cognitive functions will no longer be speculative; they will be facts. We will have decided for them what is best for them. One may argue that the simple fact of procreation is deciding what is best for future persons. Selecting a mate based on a smile, personality traits, family ties, hair color, and any other number of complex features can be positioned as deciding and choosing for future persons. The addition of procreation technology makes this choice even starker. It is important to note that this is a very different case than using CRISPR. The act of procreation in these bounds is still non-directive, in the sense that although mates can choose on largely phenotypic grounds, there is no way to direct genotypic outcomes. However, where mates can drive closer phenotype-to-genotype relationships, such as in the case of procreation techniques, those do take into account future persons:

> ...in several West European legal systems, the assumed interests of the future child are used as an important justification for restrictions on the use of procreation techniques and access to fertility treatment. (Haaf, 2016)

The use of CRISPR, in this case, would, it seems, fall closer to the category of procreation techniques and, as such, do need the consideration of an anticipatory perspective concerning future persons:

> Several areas of the law show an anticipatory form of protection of entities that may exist in the future with respect to the legal personhood it one day will have. (Haaf, 2016)

Not considering the lived experience of future persons is tantamount to discounting their experience, a reflective move that would be considered by most as unethical (Matheny, 2007). Phil Torres in "Morality, Foresight, and Human Flourishing" uses the concepts of the futurist Wendell Bell to articulate a series of ethical obligations to future generations. Among them are:

1. A concern for present people implies a concern for future people.
2. Humble ignorance ought to lead present generations to act with prudence toward the well-being of future generations.
3. The present generation's caring and sacrificing for future generations benefits not only future generations but also itself.

Earlier, we considered the imaginative reflection of each stakeholder across the retrospective/prospective process and distinguished short-term, intermediate-term, and long-term time horizons. Looking at *somatic* CRISPR modification of a genetic condition in the short-term imaginative reflection, we concluded that the treatment should be administered. Would this still be the case for the long-term time horizon?

3.7 Philosophy of Design

To answer this question, we need to use some concepts from the philosophy of technological design. While it seems to be commonly accepted that technology and technological artifacts affect agents and societies, historically, the focus of ethical analysis has been on the moral decision-making of the human actors who are the designers of technological artifacts. An anticipatory ethical analysis concerning technological design requires the consideration of at least four principles:

1. The intentions of the designer of the technology
2. The interaction and mediation of the designer's intention with the technological artifact
3. The interaction and mediation of how the user uses the technological artifact in accord with the designer's intention, in contrast with their own intentions
4. The artifact mediating between the designer's design intent and the user's use intentions and applications.

Several distinctions emerge from an examination of these perspectives. Different human actors have different kinds of control involving: how artifacts are designed, how the design exhibits the designers' intentions, how users employ artifacts according to the designers' intentions, how users formulate their intentions, and how artifacts mediate between designers and users' intentions (Noorman & Johnson, 2014).

These distinctions have opened a largely unsolved debate about whether technologies like CRISPR can intrinsically carry moral value independent of the creator's intentions. Several theorists disagree with the notion that artifacts contain moral values in themselves and support what is called the "value-neutrality thesis" (Klenk, 2020). In addition, it has been argued by Van de Poel and Kroes (2013), amongst others, that technologies such as CRISPR do contain moral values even if they are not similar to traditionally understood moral agents. This has been referred to as "value embedding for technological artifacts." Klenk (2020) has posited that this view is difficult to discern from an epistemological perspective because the end user does not have direct perceptual access to the design intentions of the original conception of the artifact. Klenk proposes the "affordance accounting of value embedding." In short, technological artifacts can have affordances as a function of response-dependent properties. In the case of CRISPR, the moral values that can be embedded are a function of its use:

> Affordances are relational properties of objects that make particular action likely, given some circumstance. Response-dependent properties are properties of objects that depend on the objects' intrinsic and extrinsic attributes. (Klenk, 2020)

Thus, in each of the "value embedding" accounts, it is foundational that technological artifacts can have, in some contexts, intrinsic moral values. From each of these perspectives in the philosophy of technological design, we can make the following assertions (Vermaas, 2010):

1. Technology and technological artifacts cannot be reduced merely to designers' intentions. Technological artifacts have contributed to developments in personal and social worlds through mediation that can go beyond the designer's intention.
2. Technology and technological artifacts actively contribute to decisions by rational and moral agents and how they make decisions to act and to the performance of these actions.
3. Technology and technological artifacts are "morally charged" (Winner, 2020).
4. Technology and technological artifacts must be designed according to principles of inter-disciplinarity and collective intentionally.

Definitions of design often begin from Herbert Simon's view that design is concerned with "changing existing situations into preferred ones." (Vermaas, 2010). To illustrate, we will use as an example the field of bioengineering. Bioengineering is an activity that is concerned with the creation of material artifacts that combine approaches from other, larger fields of science. In this example, bioengineering, medicine, and design can combine to make artifacts. These artifacts can be as diverse as neural-mediated prosthetics to novel biomaterials designed using a synthetic biology approach. In addition to being concerned with the development of the artifact, bioengineers and physicians must be concerned with how these artifacts will interact with human agents and mediate between the agent and that agent's lived experience and the lived experience of other stakeholders in the world.

Combining our assertions, we can conclude that bioengineering design is, in part, concerned with creating artifacts that will alter existing situations into more preferred ones while also considering how human actors will use these artifacts. This must include both the artifacts themselves and the human agents interacting with the artifacts. In the case of CRISPR, this requires stakeholders to consider how human agents will use and be affected by technological artifacts, which carry ethical and social ramifications.

This leads directly back to the issue of designing a technological artifact such as a gene-editing platform. Problems with technological artifacts begin to emerge when we attempt to identify what it means to design an artifact, particularly in the context of a longer time horizon. In developing a framework for assessing the longer time horizon, we can employ a set of anticipatory frames based on a technological design philosophy that involves looking at possible outcomes for various designs of the technology. To assess CRISPR in the context of a long-time horizon, and particularly in the context of the lived experiences of future persons, we need to take into consideration:

- The intended design of the artifact.
- What the artifact is and what it will be used for.
- What changes will potentially occur to the world (both individual and intersubjectively shared life worlds) because of the artifacts' use.

From this framework, several questions can next be articulated about the design aspect of CRISPR to guide an anticipatory ethical assessment for the lived experience of future persons. This list is not all-encompassing, but is designed to illustrate a potential framework for evaluation.

- Once an artifact is designed for an intended purpose and then produced, is the artifact's usability "fixed," or can the user alter the use? In this case, is the artifact the same artifact or something completely new?
- In the case of CRISPR, does this bound our ethical assessment to only the intended uses determined by the designer? The current scientific literature has articulated uses and alterations from the original intended purpose.
- Does this assessment then extend to CRISPR as the user alters it?

- Are the intentions of the designer of an artifact fixed in the artifact so that users of the artifact will somehow use the artifact according to the intentions of the intentional design of the designer, or can the users of artifacts formulate their intentions about how to use the artifact?
- Can the user's intentions and how they use the artifact alter the designers' intentions as they formulate unique perturbations based upon how they see it being used in unique ways? In the case of CRISPR, can the idea of what the original intentions were be reformulated over time as improvements on the original technology? Does this change our ethical assessment for the lived experience of future persons?

These questions are an essential starting point for understanding responsibilities and determining who is responsible when ethical issues arise. In the context of this series of intentional interrelationships and how they affect present and, more importantly, future stakeholders, anticipatory ethics needs to be considered. The intentional relationships (which can also be interpreted as affordances) are fluid. This fluidity and its effect on the lived experience of future persons is dependent on the postphenomenological concept of multistability as put forth by Ihde (2012). This refers to the capacity of any technology to be used for multiple purposes. Technologies and technological artifacts have meaning in multiple contexts and support numerous user relationships. According to Rosenberger,

> Multistability refers to the capacity of any technology to support a variety of uses and meanings. (Rosenberger, 2016)

Hammers provides a great example to help illustrate this point. In how many ways have we used hammers? Or how many different things have we used as hammers? This idea is often wielded against totalizing accounts of technology that fail to recognize the variability and context-dependency of human-technology relationships. These distinctions also hold for CRISPR gene-editing technology. Thus, artifacts can have multistable, response-dependent properties articulated as affordances. These properties are variable and context-dependent and can potentially extend beyond the designer's intentions as they are used to change existing situations into preferred situations. These select situations can affect the moral and ethical judgment of current rational and moral agents and actively contribute to decisions by rational and moral agents as to how they make decisions to act. The actions that rational and moral agents take in the present concerning multistable technologies like CRISPR can potentially have profound effects on the lived experience of future stakeholders on our long-time horizon.

Traditional ethical analyses regarding technological advancement have been approached from an instrumentalist perspective. There has been a separation of the human and technological realms. As a result of this separation, there is a failure to see that "technologies play a constitutive role in our daily lives" (Verbeek, 2012) and drive a hybrid character of agency and morality.

3.8 CRISPR and Stakeholders: Short- and Long-Term Considerations

Phenomenology, postphenomenology, and the philosophy of technological design (in conjunction with other fields such as anthropology, sociology, and psychology) have shown that there is little doubt that the world of technical artifacts exerts a powerful influence on human agents. This influence exerts effects on both the life-world and intersubjective worlds of stakeholders. In the case of CRISPR, its mere existence as an artifact with currently understood affordances biases the frameworks that current stakeholders use to assign moral values to its use in the short-term time horizon. In addition, its use continues to drive present bias, which has additional influence on the lived experience of future users of the technology.

> The moral relevance of technology is closely related to this active contribution technologies make to human practices and experiences. (Verbeek, 2012)

Different stakeholders see different possible uses for CRISPR in the present. Research scientists see the possibility of the improvement and perfection of CRISPR as a technique. Clinicians see the potential of clinical intervention for an individual with a genetic condition. A patient sees the possibility of a somatic or germline alteration (a special case that we will discuss later) that alleviates a set of symptoms arising from the genetic condition. As previously stated, it is difficult to deny that once discovered and identified, technologies and technological artifacts such as CRISPR exercise a great deal of influence upon the intersubjective experiences of humans and influence the life-world, forming the background for all the stakeholders. In these ways, the use of CRISPR influences moral frameworks around gene editing both in the short-term and long-term time horizons. The moral values surrounding the development and use of CRISPR in humans across the long-term time horizon are subject to the influence of CRISPR upon human perceptions of themselves as moral agents in the short term.

This would lead us to hypothesize that if we use CRISPR to correct a genetic condition in the short-term horizon, there would be no set, or a minimal set, of difficult ethical considerations in the future. Thus, we could conclude that because we see few significant ethical challenges with using CRISPR *now* to correct genetic conditions, there would be few (or even none as we expect CRISPR to improve as a technique) ethical quandaries over the long-term horizon concerning the use of CRISPR in humans.

However, because of the multistable nature of CRISPR and its values embedded by affordances, it follows that moral agents assigning moral conceptions in the long term will have very different moral frameworks than those agents in the short term. Research scientists, clinicians, and patients have moral thoughts in both the present and about the future and about the design and use of CRISPR, which may or may not coincide with the function for which it was originally created and designed. Additionally, morally charged technologies like CRISPR shape and condition human activities and lived human experience and practices. Given that technologies like CRISPR alter current human practices, it follows that they will also alter both the individual life-world and intersubjectively shared life-world of future humans.

Ethics is fundamentally concerned with how we are to act and how we are supposed to live. In that case, it is essential to see how technological artifacts shape our moral conceptions about these subjects in both the short and long terms. Ethical considerations about the use of CRISPR in the present may not fully capture stakeholder perspectives in the long-time horizon. Technologies mediate between us and the world and between ourselves and others. By playing a mediating role (as we have described through both Ihde's postphenomenology and ideas from the philosophy of design), technology influences the nature of our lived experience. Due to the alterations that can be made to the human genome, ethical thought needs to be focused on how we will design the future world using CRISPR gene editing. The exponential rate of technological development and innovation with technology such as CRISPR should cause a moral assessment of both consequences for present and future persons while recognizing that the current use of CRISPR has already fundamentally altered the moral framework we can use to assess the effects of CRISPR on the lived experience of future persons.

> ...the phenomenon of technological mediation lays bare how technologies also contribute to the moral actions and decisions of human beings. Technologies contribute actively to how humans do ethics. (Verbeek, 2011)

When future stakeholders perform a detailed retrospective reflection concerning the use of CRISPR as it is being employed in the present, can we hypothesize what they think about it? The answer is probably not. Therefore, we must use a framework from classical bioethics to guide us on how to think about potential future stakeholder perspectives. From this framework, we can make an educated guess as to how future persons may reflect on the effect CRISPR has on their intersubjective lived experience from a moral perspective.

3.9 Classical Bioethics

Since our ultimate goal is to analyze CRISPR concerning future stakeholders' lived experiences, we need to discuss some additional concepts from classical bioethics that can be applied to the analysis. A detailed treatment of each of these concepts is beyond the scope of this manuscript. Flynn (2021) delineates several major approaches to moral theory within classical bioethics. These approaches range from a top-down model (high moral theory), a mid-level approach, a bottom-up case-by-case approach (casuistry), and an approach referred to as reflective equilibrium. We contend that the combination of three classical bioethics approaches (mid-level, casuistry, and reflective equilibrium) can form the basis for an anticipatory ethical analysis of stakeholders' perspectives on CRISPR that consider the notion of future stakeholders' lived experiences.

The issue of what theoretical approach to take when conducting an applied ethical analysis is significant from the perspective of philosophical ethics. However, from the perspective of the stakeholders involved in real-world circumstances, the discussion of theoretical ethics suffers from the problem of not introducing stakeholders to materials, such as a wide variety of details in case studies, that are important for addressing ethical issues that could be encountered in lived experience and within personal and professional existence. For this reason, fundamental to an anticipatory stakeholder analysis is the application of casuistry to the analytic framework. Casuistry is typically referred to as a "bottom-up" approach to classical bioethics and starts from the premise that moral reasoning is casuistic in nature and, at its core, relies on analogical reasoning comparing cases as opposed to abstract reasoning about ethical quandaries (Giacomini, 2005). Because we have discussed the importance of phenomenology and postphenomenology in providing ethical guidelines, the casuistry we propose is typically referred to as "moderate casuistry" (Jonsen & Toulmin, 2012).

The comparison of historical cases with present cases needs to take into consideration technological developments and future advancements. Thus, it is crucial to recognize that this type of analysis can also be applied to projected future developments on subjects (such as CRISPR-driven interventions) and projected future cases (of CRISPR-driven interventions). Based on a comparison of cases drawn from the past and comparing these cases with present cases, we can project future cases. The development of projective future-looking cases can then form the foundation for developing an anticipatory ethical analysis. The casuistry framework is not without criticism. Flynn (2021) highlights:

> ...given its disregard for theoretically-derived principles, radical casuistry may amount to no more than a refinement of our intuitive responses to cases. It might be thought to be morally conservative, ill-equipped for social critique, with the casuist as mere expositor of pre-ordained moral norms (Arras 2017: 60–67). Finally, the approach seems to require pre-established agreement on fundamental values in order to reach conclusions, agreement that may well be lacking in pluralistic modern society.

Nonetheless, in our view, questions involving medical ethics require, at least in part, the use of case studies and other related methods to derive analogical frameworks. The analogical frameworks that flow out of a casuistic approach (specifically those analogical frameworks that are future-facing and anticipatory) can then be subject to a reflective equilibrium-driven process.

> ...reflective equilibrium is a process by which our considered responses to actual cases influence our moral principles, and those improved-upon principles then provide enhanced guidance for our response to further cases. (Flynn, 2021)

Although the emphasis here is on actual cases, an anticipatory framework allows us to develop cases based, with scientific grounding, on the future applications and advancement of technology like CRISPR. We can use analogical frameworks developed during the current case study to project anticipated developments or use of the technology. To broaden an anticipatory ethical analysis of future stakeholder perspectives concerning the use of CRISPR in humans, we can combine casuistry, reflective equilibrium, and narrative ethics with the phenomenological concepts of intersubjectivity and life-world. The postphenomenological concepts of multistability and embodiment relations can also be combined with classical phenomenology to derive an ethical perspective on the technology side of the equation (Fig. 3.3). Since we have not defined narrative ethics at this point, we now pull it into the discussion as:

> …an approach to ethics that emphasizes storytelling, the importance of the voice and perspective of the storyteller and/or patient… (Flynn, 2021)

Narrative ethics necessarily considers all aspects of stakeholders' lived experiences. Indeed, this dovetails with reflective equilibrium when applied to an anticipatory casuistic approach. Moreover, the narrative can act to wrap around and enrich the ethical analysis, particularly when hypothesizing about the lived experience of future stakeholders:

> Another way of thinking about narrative as a supplement to principles is to consider the way in which principles and cases share a dialectical relationship within the context of reflective equilibrium. Here, the cases in response to which we develop considered judgments are themselves narratives. (Flynn, 2021)

Thus, in the case of a classical bioethical analysis of all of the stakeholder-involved cases, what should again be at the center of the analysis is the *lived experience* manifest as a *future narrative* of all the stakeholders affected by gene editing. This type of ethical analysis mediated through narrative and case study should sensitize all stakeholders to this variety of lived experiences of the various stakeholders

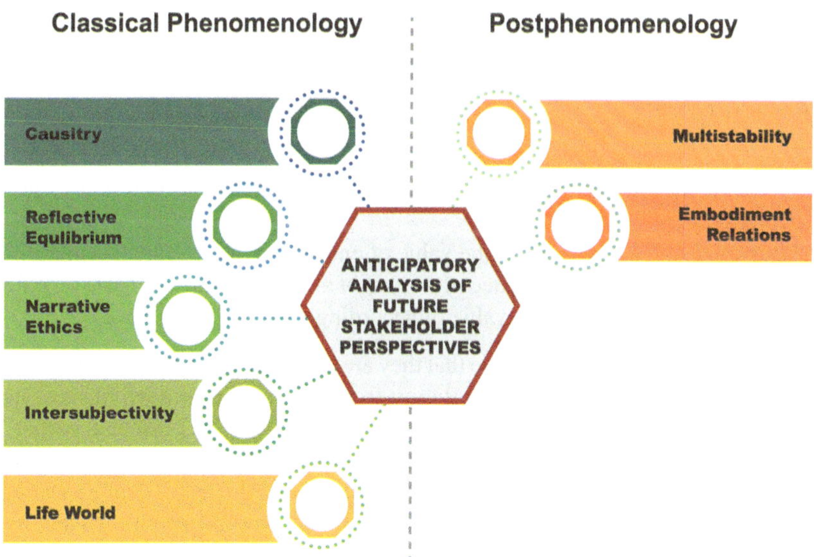

Fig. 3.3 The combination of concepts and approaches from classical phenomenology and postphenomenology can be merged with casuistry and reflective equilibrium to develop a novel anticipatory ethical analysis of CRISPR that has a basis in the lived experience

involved with, or exposed to, CRISPR gene editing. We can also combine these bottom-up approaches with "mid-level theorizing" to further develop our ethical perspective.

3.10 Anticipatory Classical Bioethics

Let us consider the following. Ethical inquiries related to CRISPR use are partially based upon what CRISPR interventions do as an action, but this action is also an extension of what a person does. In other words, the actions of CRISPR technology are also capable of being evaluated based upon the actions of the person controlling them. At the same time, the subjects upon whom the technology is used are choosing to have an action performed on them. If this is true, there are three possible levels of evaluation of the use of CRISPR. We can evaluate the actions of a person controlling the actions of CRISPR. We can evaluate the intentions of the person controlling and directing the actions of CRISPR. We can assess the consequences of the actions intended by the person controlling the actions of CRISPR. From the perspective of researchers, clinicians, and patients upon whom CRISPR is used, we can evaluate each of their intentions, their actions, and the consequences of their actions. Narratives and scenarios about these intentions and actions are built up using the methodologies that we have just outlined. How can we evaluate these scenarios using an ethical framework that can be translated into policies?

The mid-level theorizing approach built up by Beauchamp and Childress in *Principles of Biomedical Ethics* (1979) develops a structure of moral norms that allow the derivation of an applied ethics approach to wrap around other methods in classical bioethics (such as casuistry or reflective equilibrium). Historically, these norms have been the basis for standard bioethical analysis for approximately the last 40 years. These moral principles are:

1. The principle of autonomy (the value of self-direction regarding one's life and choices)
2. The principle of beneficence (the value of enhancing the welfare of others)
3. The principle of nonmaleficence (the value of avoiding imposing harm on others)
4. The principle of justice (the value of according each person her due).

These principles are so pervasive that they are reflected in the National Academies of Science Human Genome Editing Report from 2017, which we will discuss later. We can also use these moral principles to evaluate the case we presented earlier. To refresh, the case was one in which the patient is made aware, by health care professionals, of a genetic condition caused by a mutation in a gene or set of genes in the patient. This condition is not a mental condition, and we can hypothesize that it is a physiological condition such as sickle cell anemia. Without a genetic engineering intervention, the patient will suffer and potentially die from the mutation. Using our short-term/long-term time horizon distinction, we can see that in the short term, the use of CRISPR in this patient, so long as they consent to the procedure, does not violate the principles of autonomy, beneficence, nonmaleficence, and justice developed by Beauchamp and Childress. We recommended before and do again that the patient gets the treatment. But what about the lived experience of *future* stakeholders?

The use of CRISPR in the present represents an ethical dilemma concerning future stakeholders, especially when it may be used to treat neurodevelopmental, neurodegenerative, and psychiatric conditions. We will discuss this in more detail later, but it is crucial to sketch an argument now.

The current use of CRISPR is focused on somatic gene editing because of the potential for CRISPR editing in somatic cells to treat Mendelian disorders, cancer, and infectious disease (Soldner & Jaenisch, 2018). As we have highlighted in previous work, the manipulation of the human germline is an eventuality as there is a push to "correct" genetic conditions and enhance the breadth of human experience (Nestor & Wilson, 2018). Currently, mitochondrial replacement therapy in human oocytes can edit DNA mutations and has been legalized in the United Kingdom (Castro, 2016). This approach represents a bridge to CRISPR germline editing, and thus this approach is developing scientific and clinical precedent. CRISPR use in germline editing need not only be used for simple enhancement purposes or creation of "designer babies" but to confer enhanced immunity to infection and give resistance to environmental toxins (Bier & Nizet, 2021). Indeed, under the guise of HIV resistance, the landmark CRISPR edited babies were created (Cyranoski, 2019). This has led some to ask whether restrictive gene editing policies are even necessary and applicable to future uses of CRISPR (De Wert et al., 2018). Arguments have been made that genetic enhancements that are not directly related to treating inherited genetic conditions (instrumental/non-medical genetic enhancement) are not necessarily morally questionable because these enhancements will open up more possible avenues for a promising future life (Buchanan et al., 2000; Glover, 2006; Savulescu, 2001). Finally, as we will discuss later in this manuscript, there is a precedent for the combination of CRISPR with gene drive to enhance the heritability of CRISPR editing in the human germline (Nestor & Wilson, 2020).

Taken together, it is not unreasonable to hypothesize that CRISPR germline editing will continue to receive steady and growing policy support and will be bolstered by additional moral arguments for its use. This is one of the central assumptions of this manuscript. We are not arguing that our framework will stop the use of CRISPR in germline editing. As we have stated in other writing, we see it as an inevitability, albeit morally questionable, from the perspective of future persons. In some ways, our argument is not new. It has been posited that:

> …future generations have a right to an unmodified gene pool. (Frankel & Chapman, 2000).

Although this premise is also at the center of our perspective, we believe that the use of CRISPR in the short-term time horizon on individuals does not represent a significant moral issue. However, the use of CRISPR does violate three of the four Beauchamp & Childress's mid-level principles when examined from the lived experience of future stakeholders. Let us assume that CRISPR is used to correct monogenic inherited genetic conditions in the short term. Based on our earlier discussion of the postphenomenological analysis of technology, we can surmise that in the lived experiences of stakeholders in the present, the constant use of CRISPR in this context will normalize the idea of its use to correct more and more types of conditions and applications, eventually branching into germline genetics to ameliorate congenital diseases. It follows that social pressure will build towards using CRISPR for non-medical genetic enhancement. This is specifically acute for using CRISPR to ameliorate neurodevelopmental conditions that affect IQ, such as Fragile X or autism. As Harvard geneticist George Church has stated:

> If you can prevent mental retardation with gene therapy, presuming that that's permissible, then there's a whole spectrum of intellectual challenges that will be discussed. (Mahon, 2017)

This perspective is not controversial and follows from how technologies have been used in multistable ways in the past. It is very tempting to use germline CRISPR editing to, in the name of enriching of the lived experience in the present and future, go from disease amelioration to human enhancement. To this end, De Wert et al. (2018) state:

> Competition for a market share may lead…companies to exaggerate both the heritability of the relevant traits and the probability of the effectiveness of [germline gene editing]. They might also see an attractive market in advertising 'smart combinations' of [germline gene editing] and life-style modification, thus bypassing any accusations of outdated genetic reductionism.

From the perspective of future stakeholders, this use of CRISPR violates the principle of autonomy by removing the ability for future persons to direct the nature of their genetic program and thus their own lived experience

The principle of nonmaleficence is violated by potential downstream effects CRISPR may have on subsequent future individuals (beyond short-term off-target effects that currently exist in the system). We will discuss this more in our discussion of the combination of CRISPR with techniques such as gene drive. However, it is essential to state here that a significant risk precedes the need to use any additional methods—*antagonistic pleiotropy*. Antagonistic pleiotropy becomes a risk as complex multigenic conditions are treated with CRISPR. When a deleterious genetic pathway is corrected via CRISPR, there is the potential for an increased risk of enhancing other deleterious pathways that contribute to the development of other related conditions (Carter & Nguyen, 2011). This is a significant concern and one that current research does not have enough empirical data to assess.

Finally, the use of CRISPR potentially violates the principle of justice for future humans. This is in multiple ways, but specifically in the context of health disparities that currently exist. Many of these disparities are due to institutional structures that have historically discriminated against poor and minority stakeholders (Baciu et al., 2017). There is no reason to believe that future persons from these groups will be affected by the lack of current access to CRISPR technologies, specifically when it comes to non-medical enhancement.

It is essential to respond here again to the criticism that conceptualizing CRISPR in the context of enhancing complex phenotypes such as IQ is "unrealistic" or "science fiction":

> …it is important to acknowledge that any enhancement-oriented [germline gene editing] of complex traits is far beyond what may become possible in the next few years or even decades—and may never become a realistic option. (De Wert et al., 2018)

First, we should say that stakeholders do not need to engage in germline CRISPR editing to violate Beauchamp and Childress' mid-level principles. The mere use of CRISPR to alter any characteristics in a current human (including non-mental characteristics) has the potential to alter the lived experience. Based on a phenomenologically-based lived experiential approach from an ethical perspective, and that of the principle of beneficence from classical bioethics, this may not present any ethical dilemma. As we will discuss, non-germline CRISPR editing of mental conditions may represent a different ethical class. Thus, we do not need to consider germline modification to develop an anticipatory ethical analysis. However, the rapid and exponential advancement of CRISPR and the expected continual development of the technology combined with its multistability requires a projective analysis. At a bare minimum, these developments force us to respect the possibility that mental characteristics and other complex multigenic traits (like IQ and processing capacity, for example) will eventually be edited by CRISPR either at the germline or somatic levels or both.

Therefore, we are obligated to evoke a form of the precautionary principle. Simply stated, the precautionary principle emphasizes a position of caution where there is scant scientific evidence around the use of technology and its potential for harm. It has been argued that the precautionary principle can be both a trivial truism and an unscientific block to progress (Sunstein, 2001). However, the precautionary principle as formulated by the European Commission (Bourguignon, 2016) has been developed as encompassing the following set of principles:

- Proportional to the chosen level of protection
- Non-discriminatory
- Consistent with measures already taken
- Based on an examination of the benefits and costs of action or lack of action
- Subject to review in the light of new scientific data
- Capable of assigning responsibility for producing the scientific evidence necessary for a more comprehensive risk assessment

Using this set of criteria applied and applying them to the application of CRISPR and the lived experience of future stakeholders with respect to correcting multigenic conditions gives a sound grounding for an anticipatory ethical analysis that is not projecting an unrealistic application of the technology.

References

Baciu, A., et al. (2017). *Communities in action: Pathways to health equity*. The National Academies Press.

Bier, E., & Nizet, V. (2021). Driving to safety: CRISPR-based genetic approaches to reducing antibiotic resistance. *Trends in Genetics, 37*(8), 745–757. https://doi.org/10.1016/j.tig.2021.02.007

Bourguignon, D. (2016). The precautionary principle: Definitions, applications and governance – Think Tank. Retrieved from https://www.europarl.europa.eu/thinktank/en/document.html?reference=EPRS_IDA(2015)573876

Buchanan, A., Brock, D. W., Daniels, N., & Wikler, D. (2000). *From chance to choice. Genetics & justice*. Cambridge University Press.

Carel, H. (2010). Phenomenology and its application in medicine. *Theoretical Medicine and Bioethics, 32*(1), 33–46. https://doi.org/10.1007/s11017-010-9161-x

Carter, A. J. R., & Nguyen, A. Q. (2011). Antagonistic pleiotropy as a wide-spread mechanism for the maintenance of polymorphic disease alleles. *BMC Medical Genetics, 12*, 160.

Castro, R. J. (2016). Mitochondrial replacement therapy: The UK and US regulatory landscapes. *Journal of Law and the Biosciences, 3*(3), 726–735. https://doi.org/10.1093/jlb/lsw051

Cyranoski, D. (2019). The CRISPR-baby scandal: What's next for human gene-editing. *Nature, 566*(7745), 440–442. https://doi.org/10.1038/d41586-019-00673-1

De Wert, G., Heindryckx, B., Pennings, G., Clarke, A., Eichenlaub-Ritter, U., Van El, C. G.,... Cornel, M. C. (2018). Responsible innovation in human germline gene editing: Background document to the recommendations of ESHG and eshre. *European Journal of Human Genetics, 26*(4), 450–470. https://doi.org/10.1038/s41431-017-0077-z

Drummond, M. (2015). *Methods for the economic evaluation of health care programmes*. Oxford University Press.

Drummond, J. J. (2010). *The A to Z of Husserl's philosophy*. Lanham: Scarecrow Press.

Flynn, J. (2021). *Theory and Bioethics, The Stanford Encyclopedia of Philosophy*, Edward N. Zalta (ed.), https://plato.stanford.edu/archives/spr2021/entries/theory-bioethics

Frankel, M. S., & Chapman, A. (2000). Human inheritable genetic modifications: Assessing scientific, ethical, religious and policy issues. https://www.aaas.org/sites/default/files/migrate/uploads/germline.pdf

Gallagher, S. (2012). *Phenomenology. Houndmills, Basingstoke*, England: Palgrave Macmillan.

Giacomini, M. (2005). One of these things is not like the others: The idea of precedence in health technology assessment and coverage decisions. *The Milbank Quarterly, 83*(2), 193–223. https://doi.org/10.1111/j.1468-0009.2005.00344.x

Glover, J. (2006). *Choosing children. The ethical dilemmas of genetic intervention*. Clarendon Press.

Haaf, L. T. (2016). Future persons and legal persons: The problematic representation of the future child in the regulation of reproduction. *Laws, 5*(1), 10. https://doi.org/10.3390/laws5010010

Husserl E (1970) *The Crisis of European Sciences: An Introduction to Phenomenological Philosophy*. (translated by: Carr D) Northwestern University: Evanston.

Husserl E (1980) *Ideas Pertaining to a Pure Phenomenology and to a Phenomenological Philosophy*, Third Book. Phenomenology and the Foundation of the Sciences. (Translated by Klein T and Pohl W) Kluwer: Dordrecht.

Husserl E (1990) *On the Phenomenology of the Consciousness of Internal Time* (1893–1917). (translated by John Brough) Kluwer: Dordrecht.

Ihde, D. (1990) *Technology and the Lifeworld: From Garden to Earth*, Indiana University Press, Bloomington.

Ihde, D. (2012). *Experimental phenomenology: Multistabilities*. Albany: State University of New York Press.

Irwin, S. O. (2017). *Digital media: Human-technology connection*. Lexington Books.

Jacobs, H. (Ed.). (2021). *The Husserlian Mind* (1st ed.). Routledge. https://doi.org/10.4324/9780429243790

Jonsen, A. R., & Toulmin, S. E. (2012). *The abuse of casuistry: A history of moral reasoning*. University of California Press.

Klenk, M. (2020). How do technological artefacts embody moral values? *Philosophy & Technology*. https://doi.org/10.1007/s13347-020-00401-y

Mahon, P. (2017). *Posthumanism: A guide for the perplexed*. Bloomsbury.

Mapp, T. (2008). Understanding phenomenology: The lived experience. *British Journal of Midwifery, 16*(5), 308–311. https://doi.org/10.12968/bjom.2008.16.5.29192

Matheny, J. G. (2007), Reducing the risk of human extinction. *Risk Analysis 27*, 1335–1344. https://doi.org/10.1111/j.1539-6924.2007.00960.x

Moran, D., Cohen, J. D., & Cohen, J. (2012). *Husserl dictionary*. Bloomsbury Academic.

Nestor, M. W., & Wilson, R. L. (2018). Can the use of CRISPR in humans result in decreased social justice for future stakeholders? *Ethics in Biology, Engineering and Medicine: An International Journal, 9*(1), 5–16. https://doi.org/10.1615/ethicsbiologyengmed.2019030551

Nestor, M. W., & Wilson, R.L.(2020). Beyond Mendelian Genetics: Anticipatory Biomedical Ethics and Policy Implications for the Use of CRISPR Together with Gene Drive in Humans. *Bioethical Inquiry. 17*, 133–144. https://doi.org/10.1007/s11673-019-09957-7

Noorman, M., & Johnson, D. G. (2014). Negotiating autonomy and responsibility in military robots. *Ethics and Information Technology, 16*(1), 51–62. https://doi.org/10.1007/s10676-013-9335-0

Poel, I. V., & Kroes, P. (2013). Can technology embody values? In *Philosophy of engineering and technology the moral status of technical artefacts* (pp. 103–124). https://doi.org/10.1007/978-94-007-7914-3_7

Rosenberger, R., & Verbeek, P. P. C. C. (2015). *A field guide to postphenomenology*. In R. Rosenberger, & P-P. Verbeek (Eds.), Postphenomenological Investigations: Essays on Human-Technology Relations (pp. 9–41). (Postphenomenology and the Philosophy of Technology). Lexington Books.

Rosenberger, R. (2016). Husserl's missing multistability. *Techné: Research in Philosophy and Technology, 20*(2), 153–167. https://doi.org/10.5840/techne20168356

Savulescu, J. (2001). Procreative beneficence: Why we should select the best children. *Bioethics, 15*(5–6), 413–426. https://doi.org/10.1111/1467-8519.00251

Schwab, K. (2017). *The fourth industrial revolution*. Currency.

Smith, D. W. (2013, December 16). Phenomenology. Retrieved from https://plato.stanford.edu/archives/sum2018/entries/phenomenology/

Soldner, F., & Jaenisch, R. (2018). Stem cells, genome editing, and the path to translational medicine. *Cell, 175*(3), 615–632. https://doi.org/10.1016/j.cell.2018.09.010

Stanghellini, G., & Mancini, M. (2017). The *therapeutic interview in mental health: A values-based and person-centered approach*. Cambridge University Press. https://doi.org/10.1017/9781316181973

Sunstein, C. R. (2001). *The laws of fear*. Law School, the University of Chicago.

Svenaeus, F. (2018). *Phenomenological bioethics: Medical technologies, human suffering, and the meaning of being alive*. Routledge.

Toombs, S. K. (2001). *Handbook of phenomenology and medicine*. Kluwer.

Verbeek, P. P. (2011). *Moralizing technology: Understanding and designing the morality of things*. The University of Chicago Press.

Verbeek, M. (2012). *A Guide to Modern Econometrics*," 4th Edition, Wiley, New York

Vermaas, P. E. (2010). *Philosophy and design: From engineering to architecture*. Springer.

Winner, L. (2020). Do artifacts have politics? In *Emerging technologies: Ethics, law and governance* (pp. 15–30). https://doi.org/10.4324/9781003074960-3

Zaner, R. M. (1981). *The context of self: A phenomenological inquiry using medicine as a clue*. Ohio University Press.

Chapter 4
CRISPR and the Concept of Personhood

4.1 Introduction

We have established that a discussion of the use of CRISPR to treat multigenic conditions is not dismissible on the grounds that it is science fiction or unrealistic, although not scientifically possible at the writing of this manuscript. We have also established why an anticipatory approach to ethics is important for keeping pace or even front-running rapidly developing technologies. Concepts from the phenomenological philosophy of Husserl were introduced to guide our ethical analysis, which focused on the lived experience of three exemplary stakeholders affected by CRISPR. In this chapter, we discuss a variety of ways personhood could be affected by CRISPR-mediated interventions. At the center of this discussion is the definition of personhood in the lived experience of stakeholders affected by CRISPR and the associated residual effects related to the life-world.

It is now time to turn to the special case of CRISPR editing of genes that affect human cognition and perception. An astute reader at this point would ask why we are separating this category of gene editing from others. In the example of sickle cell anemia, is it not the case that using CRISPR to ameliorate the disease changes the lived experience of the stakeholder and future stakeholders? This is an essential and central question for the main premise of the ethical analysis in this manuscript. We will attempt to address this by revising the concept of lived experience.

4.2 CRISPR Editing of Complex Multigenic Traits Like Cognition

It would seem from a scientific perspective that any potentially slippery-slope concerns about CRISPR editing complex multigenic traits that affect cognition can be dismissed out of hand. This is because there is scientific evidence that genetic

M. W. Nestor, R. L. Wilson, *Anticipatory Ethics and The Use of CRISPR in Humans*, https://doi.org/10.1007/978-3-030-98368-0_4

determinism concerning cognition is scant. Cognition is a complex process that involves a dynamic interplay between gene expression at the subcellular cellular, network, gross brain area levels, and the environment. This is borne out by studies that have identified over 350 genes that underlie cognition and cognitive ability (Lam et al., 2017). To add to the complexity, epigenetics, which is in some sense a genetic reflection of the environment, can affect gene expression itself and potentially modulate cognition and perception (Bonduriansky & Day, 2018; Day & Sweatt, 2011). This can happen as early as the prenatal environment concerning fetal exposure to chemical stressors (Palma-Gudiel & Fañanás, 2021). Currently, the best that neuroscience can do is generate highly dimensional data that creates relatively low-resolution spatial maps between gene expression and functional activation in cognitive processes (Hansen et al., 2021). This is not even to mention the highly specific structure and function of individual neurons, in which dynamic expression of single-nucleotide variants (SNVs) occurs (Marshall & Bredy, 2016). These neurons can express up to 1500 somatic SNVs each and may express nearly 10,000 SNVs over a lifespan (Lodato et al., 2015; Hazen et al., 2016). In looking to treat diseases that affect human cognition, like Alzheimer's or autism, the current state of the art is to use transcriptome-wide analysis or other transcriptomic approaches combined with genome-wide association studies to identify lists of genes that may underlie difficulties with cognition (Fan et al., 2020; DeRosa et al., 2018).

These lists still need to be functionally validated to understand the relationship between genetics, structure, function, and cognition at the cellular and network levels (Lunden et al., 2020; Mlecnik et al., 2018; Díaz-Santiago et al., 2020). Many individual genes or proteins have been suggested as "master regulators" for brain signaling or cognitive function, such as REST/NRSF (Hwang & Zukin, 2018), HDAC6 (LoPresti 2020), Phosphodiesterase 4B (Tibbo & Baillie, 2020); PGC-1α (Halling & Pilegaard, 2020) and many more. The scientific case against considering CRISPR in the context of cognition and cognitive ability is supported by a 1.1-million-person genome-wide association study in which 1271 single nucleotide polymorphisms were identified. When a polygenic phenotypic analysis was performed on cognitive phenotypes and educational attainment and cognitive performance, polygenic scores only accounted for 11–13% of the former and 7–10% of the latter (Lee et al., 2018). Taken together, it seems as though cognition itself is such a complex trait that it is beyond the reach of CRISPR, at least for the foreseeable future. Then why devote time and space in this manuscript to address this topic?

Because this manuscript is focused on an *anticipatory* perspective, it is essential to explore the potential use of CRISPR to alter cognition long before it becomes a scientific reality. This, as we discussed in Chap. 1, is the essence of anticipatory ethics. It is not a scientific leap to surmise that once monogenic conditions like sickle cell disease are conquered using CRISPR approaches (and we believe they will be), developers will turn their attention to polygenic conditions and eventually focus on cognition. Here it bears mentioning again that we are not advancing a position that CRISPR in all cases is unethical but suggesting that we engage in a deep stakeholder conversation around medical and non-medical enhancement for cognition. There is already a precedent for an application of CRISPR to enhance cognition. An analogue for a potential use of CRISPR to enhance cognitive abilities can be seen in the use of chemicals to enhance cognition. The use of cognition-enhancing drugs was surveyed by Maier et al. (2018), and Lavazza (2018) summarized the findings as such:

> The highest usage rate was reported in the USA: in 2017, almost 30% of respondents said they had used smart drugs at least once in the previous 12 months, compared to 20% in 2015. But the largest increases were recorded in Europe: from 2015 to 2017, use in France rose from 3 to 16%, in the UK from 5 to 23%, in the Netherlands from 10 to 24%, in Ireland from 4 to 18%. There is an increasing demand for performance enhancement due to molecules not specifically targeted for that purpose…Therefore, the prospect of permanent and irreversible enhancement through genetic editing could replace other forms of empowerment once it becomes sufficiently safe and economically accessible.

Alteration of cognition seems like a logical place to use CRISPR in the context of treating mental conditions. For example, what neuroscientist would not want to use CRISPR to relieve the suffering patient of schizophrenia or depression? We have argued in Nestor & Wilson (2018) that:

> From a scientific perspective, some research has shown that one of the largest contributors to unhappiness is mental illness. A recent analysis from Layard et al., (2013) states, "Mental health is the biggest single predictor of life-satisfaction". Therefore, eradication of genes that play a role in depression, for example, via the use of CRISPR would increase happiness and health and lead to a better lived experience for future persons. Indeed, Lyubomirsky et al., (2005) suggests that mental states that induce happiness are at least 50% genetically determined. Therefore, at minimum, the use of CRISPR to eradicate genes that result in mental illness or depression would result in an approximate increase in happiness by 50% in future persons. At this point in the argument, we are morally compelled to use CRISPR technology in humans.

Although a simplistic example that does not highlight the complexity of cognition and mental conditions, as we highlighted earlier in this chapter, this example shows the potential draw for scientists, clinicians, and patients with respect to developing CRISPR interventions for cognition. Why, then, should this pose a problem for the lived experience of stakeholders? Because central to the lived experience is the concept of personhood. Personhood is fundamental to the lived experience and intersubjectivity because personhood extends beyond the person as:

> ...a standing or status that is bestowed upon one human being, by others, in the context of relationship and social being. (Kitwood, 1997)

How will the concept of personhood be affected by CRISPR? This depends upon the definition we give of personhood. Seay and Nuccetelli (2017) state:

> A person typically: Has preferences, conscious, desires, feelings, thoughts, a sense of time. Is able to remember its own past actions and mental states, envision a future, experience pleasure and pain, take moral considerations into account in choosing between possible actions. Is self conscious, capable of rational thought, and/or capable of rational deliberation.

4.3 CRISPR and Definitions of Personhood

It is interesting here to briefly survey various definitions of personhood from the enlightenment period to the present. (Many of these are borrowed from the excellent article on the subject of personhood in neuroscience by Farah & Heberlein, 2007)

Locke:	"...an intelligent being that has reason and reflection, and can consider itself the same thinking being in different times and places..."
Kant:	"...rational beings are called persons in as much as their nature already marks them out as ends in themselves..."
Tooley:	"...possesses the concept of a self as a continuing subject of experiences and other mental states, and believes that it is itself such a continuing entity..."
Dennett:	"...persons are rational beings...beings to which states of consciousness are attributed...capable of verbal communication...distinguishable from other entities by being conscious..."
Fletcher:	"...Below IQ 40 individuals might not be persons; below IQ 20 they are definitely not persons..."
Feinberg:	"...persons are those beings who are conscious, have a concept and awareness of themselves, are capable of experiencing emotions, can reason and acquire understanding..."
Englehardt:	"...capacity to be self-conscious, rational, and concerned with worthiness of blame or praise..."
Rorty:	"...capable of interacting with others, in a common world..."

All of these concepts of personhood have at their central premise the idea that cognition and cognitive capacity are the fundamental and necessary conditions for personhood (Farah & Heberlein, 2007). Thus, most philosophers and neuroscientists developing the concept put forth these various states/attributes of cognition as fundamental conditions of personhood. This is because of cognition's unique relation to the idea of autonomy. To assign moral values, an autonomous individual must impose a moral framework onto an action or item. Because cognition is an inherent property of persons, cognitive attributes buttress autonomy and set the stage for moral frameworks and ethical judgments (Foster & Herring, 2017). Cognition is inherent to persons, so too then is morality, according to this perspective.

It follows then that altering cognition using CRISPR will necessarily change our definition of personhood. How might that be the case? How does using CRISPR in this way change personhood at all? If personhood is mainly concerned with IQ, verbal communication, or rationality, are we not enhancing the experience and/or definition of personhood if we enhance these abilities via CRISPR? CRISPR, in this way, could be used to equalize personhood by "smoothing" variations in cognitive capacity. To wit, Foster and Herring (2017) write:

> The intellectual capacity approach appears to undermine the principle of equality (namely that all human beings are equal) which is central to the appeal of personhood. If what is of moral value are the mental capabilities specified, then inevitably some people will exhibit

> these to a greater degree than others. Some people have high states of self-awareness, are more rational, or can exercise more richly (or at least more explicitly) autonomous decisions than others. If such qualities are what generate moral value, how can we avoid the conclusion that those people are of higher moral value than others? One answer to this question is that what is of moral value is the *ability* to engage in these mental activities. On such a view the frequency or extent to which someone exercises these abilities does not matter.

It would seem that enhancing cognitive qualities using CRISPR potentially would enhance an individual's ability to engage in mental activities that generate moral value. Why then would we consider the use of CRISPR to alter cognition a unique dilemma for the concept of personhood?

To answer this question, we must separate personhood from an instrumentalist-reductionist perspective that, in the modern era, has some roots in genetic determinism. Most of the conceptualizations of personhood just discussed the trend toward reductionism. To capture a more holistic conceptualization of personhood, it is essential to integrate the definition put forth by Rorty and Kitwood, which rests on, according to Rorty, "…interacting with others, in a common world."

These intersubjective interactions rely on three main types of reasoning: theoretical, practical, and moral. Theoretical reasoning develops ideas related to how we theorize about what we may be planning on doing. In contrast, practical reasoning engages in reflecting on and determining how to act in conjunction with theoretical reasoning. Moral reasoning aims at employing 'principles' to show how moral agents are morally responsible. Moral agents can be morally accountable based on character traits, intentions, actions, and outcomes produced by those same character traits, intentions, and actions. Agency itself "is both a necessary and sufficient condition" for personhood (Wallace, 2000). In this sense, moral agency and individual ethical perspectives flow not just from the inherent property of possessing cognitive abilities but from personhood itself, which relies in part on the ability of the individual to assign value. An individual's capacity to value may be altered by altering cognition via CRISPR. This, in turn, may alter moral reasoning itself and remove the ability for future persons to have an intersubjective lived experience that relies on common moral frameworks that have an intersubjective historical grounding. Because the capacity to value may be altered in this way via CRISPR, moral grounding based on frameworks that have been used to assign moral value historically (before the use of CRISPR) may not be compatible with the lived experience of future persons who have been genetically altered via CRISPR. This, in turn, may alter the desire-belief relationships, which form a large part of the relational value of personhood and ethics (Sinhababu, 2013). In addition, according to Rorty, Kitwood, and others, personhood itself requires relational intersubjectivity. According to Foster and Herring (2017):

> People are in their very nature interdependent and vulnerable. It follows from this that the basic moral value of humans is not found in their individual capabilities or in their membership of the species, but rather in their relationships. Thus the question "is X a person?" is problematic because we can only conceive of X in the context of their relationships. We can say that X and Y are people if their relationship reveals the moral qualities that we look for in human relationships. But we cannot imagine an isolated person, and assess their

> capabilities, since such a person does not exist. It is people's relationships, rather than any inherent characteristics of a person, which have moral value and are deserving of special moral status.

Using CRISPR to alter cognitive capacity, if it could be done, would change our concept of personhood in the present and, consequently, may alter the intersubjective lived experience of future stakeholders by fundamentally altering the *interdependence* of those future stakeholders. This may occur by subtle alterations to the concept of personhood or potentially removing the idea of personhood itself either from a cognitive or relational perspective for future persons. It may remove any historical notions of morality concerning personhood, and in this way, it may prevent future stakeholders from "...interacting with others, in a common world." The common word, which can be thought of as the intersubjective experiences of the larger life-world of the stakeholders in our analysis, must rest on the lived experiences of those stakeholders.

According to Husserl, an essential question within the context of lived experience is focused on what it means to be a person. The idea of personhood is aligned with our earlier discussion in that the meaning of personhood is bounded by relations to others in the intersubjective/common life-world, sometimes referred to as a "community of monads":

> The discussion of 'persons' is introduced in 'Philosophy as a Rigorous Science' and treated in Ideas II Sec 49-51. , Cartesian Meditations Sec 32 and in his writings on intersubjectivity. Persons are members of social groupings (e.g. families) and communities, and are correlated with an environment or surrounding world (Umwelt, Ideas II sec 50). Persons approach primarily through (Ideas II sec 34) that considers human beings in terms of their inner, subjective, mental life and motivations. Persons stand in relation to values and in mutual understanding (through empathy) and communication with one another in what Husserl calls the community of monads. (Moran et al., 2012)

Lived experience is characterized by our intersubjective interactions with others in the surrounding world. As we (based on this discussion) define it, personhood is not simply cognitive ability but the social relationships we have with others in the common lived world. By altering these relationships, CRISPR can change what personhood means and potentially destabilize it if the use of CRISPR destabilizes relationships with others.

A significant question is: to what degree would using CRISPR to edit genes that mediate cognition have on each stakeholder's lived experience concerning personhood? In other words, the use of CRISPR to alter genes that mediate cognition may redefine social groupings and thus transcendental intersubjectivity itself. By changing the idea of the intersubjective life-world, the use of CRISPR to alter genes related to cognition can redefine the very nature of personhood itself. Two examples can be used to illustrate how the use of CRISPR, via its effect on our concept of personhood, may potentially diminish the intersubjective interactions between stakeholders in a common world.

To understand why CRISPR-mediated alterations of genes related to cognition (herein referred to simply as CRISPR in this chapter) in their destabilization of personhood lead to the dissolution of the stakeholder's intersubjective common world, and thus dissolve common ethics for current and future persons, we first need to connect moral agency with lived experience.

4.4 Belief–Desire–Intention Model of Agency

In order to engage in meaningful ethical relationships that are based on, and constructed from, a common intersubjective world, stakeholders in our running example need the framework of personhood. Personhood, which derives meaning from the relationships within and between each stakeholder's lived experiences, allows each stakeholder to assign moral agency and construct common ethics that can be used to develop an anticipatory ethical analysis for current and future persons concerning the use of CRISPR.

Although Husserl's phenomenology begins from a first-person account of lived experience, the ability of the various stakeholders to understand one another requires each stakeholder's ability to *mind read* and attribute mental states to one another. This apperception of other stakeholders' mental states is perceived in a specific way. Mind reading is required to develop an experiential and intuitive innate understanding of others' behaviors and perceptions and flows from a "folk psychology"; this folk psychology recursively allows for the development of what is called the "theory-theory" (Carruthers & Smith, 2010). This theory–theory is the framework by which experiences within the common lived-world drive the development and revision of theories (including assignment of ethical agency) about objects and people within each stakeholder's individual lived-world. This type of mindreading is our common lived experience of the theory of mind:

> Theory of mind…is narrowly understood as the ability to attribute beliefs and desires to others in order to predict their behaviour…Often theory of mind is taken to be an implicit theory that we unconsciously use, like the rules of grammar that shape our use of language but which we do not have direct introspective access to. (Andrews, 2014)

By destabilizing personhood, CRISPR may diminish the ability of each stakeholder to mind read, undermine folk psychology and, ultimately, interfere with the development of a robust theory of mind. We perceive other stakeholders, and each stakeholder is experienced as having their own lived experiences and inner mental life. In this way, one patient can mind read another patient's lived experience to understand their lived experience. The same thing holds true for each of the stakeholders in relation to imaginatively mindreading each other's mental lives. The patient of gene editing can mindread the clinician's lived experience and understand their perspective on the use of CRISPR. The researcher, in turn, can mindread the patient and be driven to develop more uses of the technology to serve the patient and the clinician, whom the researcher can also mindread. All these interactions are multidirectional and potentially complex. In this oversimplified example, what individuals, clinicians, and researchers have in common is a goal-oriented picture of human agency for the use of CRISPR. But how is moral agency assigned to each stakeholder from each other stakeholder concerning personhood for the patient, clinician, and researcher?

One framework that allows for the development of moral agency from the concept of personhood with respect to these stakeholders is Michael Bratman's belief–desire–intention model (BDI) :

> Belief, desire, and intention (which we will call "BDI entities") merit careful investigation because they are of paramount importance for an ontology of mental reality and agency. Inspired by Bratman's philosophy, for instance, the BDI model of agency recognizes the primacy of the BDI entities in practical reasoning and rational actions. (Toyoshima et al., 2020)

BDI requires the concept of personhood to contextualize moral behavior, particularly for belief-desire reasoning defined as such:

> The process by which one explains and predicts another's behavior on the basis of one's understanding of the other's desires and beliefs. Belief–desire reasoning is the basis for theory of mind. (APA Dictionary of Psychology, n.d.)

The BDI model plays a vital role in the theory of action and theory of agency. While the BDI model can be used to try to predict the behaviors, it can also show the structure of items one can employ to clarify one's own behavior. We assume actors and rational agents have beliefs and desires, formulate intentions, perform actions, and produce outcomes. Similarly, when we identify goals, each of us, through reflection, realizes that we have intentions, and we act upon these intentions to achieve our goals. This model is also related to rational agency (Fig. 4.1).

The BDI model plays an important role in predicting how agents will act, helping to predict how future-directed reasoning will occur in practical reasoning directed towards achieving future goals. This model can be used to predict the behavior of others while at the same recognizing an application of the same the model to ourselves. First, the model can clarify how we develop our plans to attempt to achieve our goals. These beliefs and desires are the basis of intentions, actions, and goals. The assumption made here is that this can be done with each of the stakeholder perspectives listed above. The BDI model can be used to attribute cause and effect. In this case, intentions cause actions, and actions cause outcomes.

4.5 Personhood, the BDI Model, Moral Agency, and Ethics

We next use these distinctions in conjunction with ideas about ethics developed by Dwight Furrow. Understanding the ethical issues involved with CRISPR developments requires an understanding of ethics. Ethics is taken to relate to agents who perform actions. Furrow moves beyond this characterization when he identifies the focus of ethical analysis as involving series of factors. As Furrow states, ethics is related to evaluating actions, and actions are performed by those capable of being moral agents. What Furrow means by an action goes beyond an isolated action:

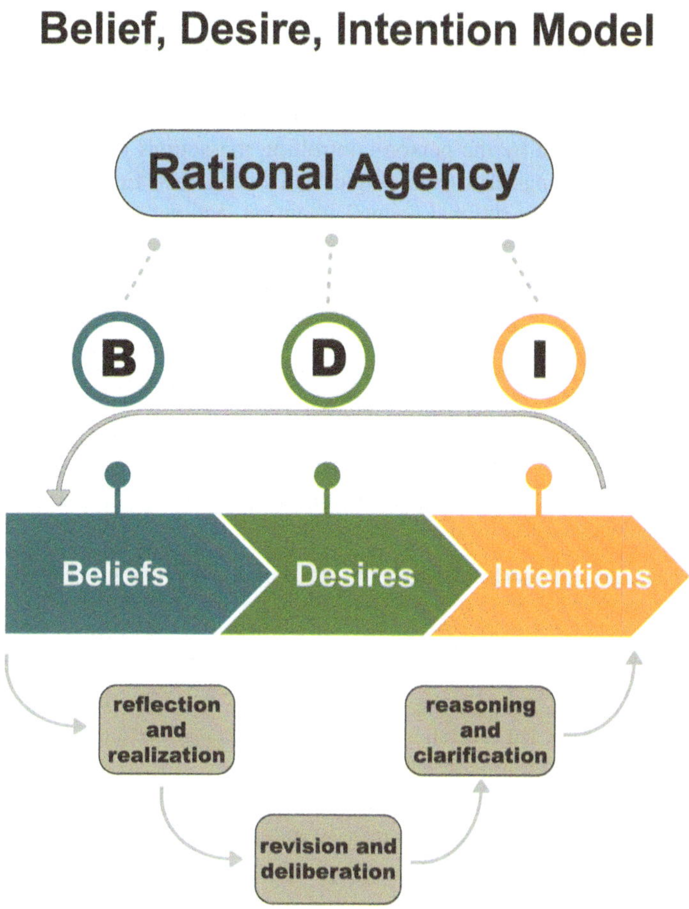

Fig. 4.1 The belief, desire, intention (BDI) model is fundamental to a theory of agency and ascribing meaning and predictive validity to actors and other rational agents. Importantly, it can be combined with reflection to understand the lived experiences of diverse stakeholders

> When we evaluate an action, we can focus on various dimensions of the action. We can evaluate the person who is acting, the intention or motive of the person acting, the nature of the act itself, or the consequences. (Furrow, 2005)

This evaluation must rely on a relational and intersubjective definition of personhood. Without the relationships to contextualize the action amongst each stakeholder's diverse lived experiences and life-worlds, it is difficult to evaluate the moral agency of each stakeholder. This framework can be applied to the application of CRISPR to alter genes concerning cognition. Ethical issues related to CRISPR are based on the idea that those engaged in using CRISPR technology are performing actions and that those actions are an extension of a person's intentions. The actions involved with using CRISPR can be evaluated based upon the intentions and actions of the stakeholders engaged in the activity. If the distinctions are applied to using

CRISPR, there are three possible levels of ethical evaluation. We can evaluate the *actions* of a stakeholder controlling the technology using CRISPR to edit genes, the *intentions* of the person controlling and directing the actions involved using CRISPR to develop short- and long-term technological advancements, and the *consequences* of the actions intended by the person controlling the actions of using CRISPR to develop short- and long-term technological advancements. Ethical issues can be identified by referencing rational agents' intention, actions, and outcomes. These generic distinctions related to moral responsibility can also be related to standard ethical principles, as we described for Beauchamp and Childress in Chap. 3.

As we stated earlier, theoretical reasoning develops ideas related to theorizing about what we may be planning on doing, while practical reasoning determines how to act. Moral reasoning then shows how moral agents are morally responsible at several levels for practical activities and actions due to beliefs, desires, intentions, actions, and outcomes. These are partly produced by the context of relational personhood in which each agent is enmeshed. This context of relational personhood (which can also be thought of as an aspect of intersubjectivity) is best illustrated by the Ring Theory of Personhood:

> The Ring Theory of Personhood suggests that personhood comprises three closely linked, dynamic areas: an innermost ring, the 'Innate Ring', is anchored in the notion that all humans are conferred with personhood as a result of being human irrespective of their stage of life; the middle ring, the 'Individual Ring', encompasses the values, beliefs, roles, personality traits, goals and preferences that a patient holds and expresses, which revolves around consciousness; and finally, the outermost ring, the 'Relational Ring', consists of a patient's personal ties and connections with others. Taken together, these three rings are what constitute a patient's personhood. (Krishna, 2013)

Ultimately, it is the *person* or *persons* controlling the use of CRISPR to develop short- and long-term technological advances that are subject to moral evaluation. If we want to identify the ethical issues with disseminating CRISPR gene-editing technology, we need to ask

- What beliefs are held about CRISPR by persons using this technology?
- What are the desires of each *person* concerning the use CRISPR?
- What are the intentions of *persons* using CRISPR?
- What actions are performed by *persons* when the CRISPR is used?
- What are the consequences for *persons* of the CRISPR gene-editing technology?
- How do all of these frameworks change depending on each person's lived experience?

It should be clear now that person is synonymous with stakeholder in our examples above. To assess the moral responsibility of stakeholders, we have to use a moral framework that shows the interrelationships between these questions for each stakeholder. We showed how BDI is a framework that can be combined with an anticipatory ethics to buttress a theory of mind that allows for mindreading and theory of action for each stakeholder. This mind reading between stakeholders will enable them to assess BDIs for each other and build a model that intersects with traditional moral theories and systems employed to assess moral responsibility for present persons concerning CRISPR. Because stakeholder moral perspectives reflect individual lived experience, to mind read, stakeholders need a common intersubjective world that depends on a common definition of personhood. If CRISPR is used to alter cognition, but only in some persons and not others, then stakeholders lose the relational ring (collapsing the ring theory of personhood), and thus personhood as we have defined it, becomes destabilized. In this case, the BDI framework would not rely on a foundational theory of mind and could not accurately assign moral agency. Using CRISPR in this way, we risk dismantling common intersubjective moral frameworks like the BDI framework and destabilizing moral agency. This leaves us without a stable, ethical framework for future persons and potentially permanently alters personhood itself.

References

Andrews, K. (2014). Personhood, ethics, and animal cognition: Situating animals in hares two-level utilitarianism, by Gary E. Varner * The philosophy of animal minds, edited by Robert W. Lurz. *Mind, 123*(491), 959–966. https://doi.org/10.1093/mind/fzu128

APA Dictionary of Psychology. (n.d.). Retrieved from https://dictionary.apa.org/belief-desire-reasoning

Bonduriansky, R., & Day, T. (2018). *Extended heredity: A new understanding of inheritance and evolution*. Princeton University Press.

Carruthers, P., & Smith, P. K. (2010). *Theories of theories of mind*. Cambridge University Press.

Day, J., & Sweatt, J. (2011). Epigenetic mechanisms in cognition. *Neuron, 70*(5), 813–829. https://doi.org/10.1016/j.neuron.2011.05.019

Derosa, B. A., Hokayem, J. E., Artimovich, E., Garcia-Serje, C., Phillips, A. W., Booven, D. V., … Dykxhoorn, D. M. (2018). Convergent pathways in idiopathic autism revealed by time course transcriptomic analysis of patient-derived neurons. *Scientific Reports, 8*(1). https://doi.org/10.1038/s41598-018-26495-1

Díaz-Santiago, E., Jabato, F. M., Rojano, E., Seoane, P., Pazos, F., Perkins, J. R., & Ranea, J. A. (2020). Phenotype-genotype comorbidity analysis of patients with rare disorders provides insight into their pathological and molecular bases. *PLoS Genetics, 16*(10). https://doi.org/10.1371/journal.pgen.1009054

Fan, C., Chen, K., Zhou, J., Wong, P., He, D., Huang, Y., … Zhao, H. (2020). Systematic analysis to identify transcriptome-wide dysregulation of Alzheimer's disease in genes and isoforms. *Human Genetics, 140*(4), 609–623. https://doi.org/10.1007/s00439-020-02230-7

Farah, M. J., & Heberlein, A. S. (2007). Personhood and neuroscience: Naturalizing or nihilating? *The American Journal of Bioethics, 7*(1), 37–48. https://doi.org/10.1080/15265160601064199

Foster, C., & Herring, J. (2017). *Identity, personhood and the law*. Springer.

Furrow, D. (2005). *Ethics: Key concepts in philosophy*. Continuum.

Halling, J. F., & Pilegaard, H. (2020). PGC-1α-mediated regulation of mitochondrial function and physiological implications. *Applied Physiology, Nutrition, and Metabolism, 45*(9), 927–936. https://doi.org/10.1139/apnm-2020-0005

Hansen, J. Y., Markello, R. D., Vogel, J. W., et al. (2021). Mapping gene transcription and neurocognition across human neocortex. *Nature Human Behaviour*. https://doi.org/10.1038/s41562-021-01082-z

Hazen, J. L., Faust, G. G., Rodriguez, A. R., Ferguson, W. C., Shumilina, S., Clark, R. A., Boland, M. J., et al. (2016). The complete genome sequences, unique mutational spectra, and developmental potency of adult neurons revealed by cloning. *Neuron, 89*, 1223–1236.

Hwang, J., & Zukin, R. S. (2018). REST, a master transcriptional regulator in neurodegenerative disease. *Current Opinion in Neurobiology, 48*, 193–200. https://doi.org/10.1016/j.conb.2017.12.008

Kitwood, T. (1997). *Dementia reconsidered: The person comes first*. Open University Press.

Krishna, L. K. (2013). Accounting for personhood in palliative sedation: The ring theory of personhood. *Medical Humanities, 40*(1), 17–21. https://doi.org/10.1136/medhum-2013-010368

Lam, M., Trampush, J. W., Yu, J., et al. (2017). Large-scale cognitive GWAS meta-analysis reveals tissue-specific neural expression and potential nootropic drug targets. *Cell Reports, 21*(9), 2597–2613.

Lavazza, A. (2018). Cognitive enhancement through genetic editing: A new frontier to explore (and to regulate)? *Journal of Cognitive Enhancement, 2*(4), 388–396. https://doi.org/10.1007/s41465-018-0104-1

Lee, J. J., Wedow, R., Okbay, A., Kong, E., Maghzian, O., Zacher, M., Nguyen-Viet, T. A., Bowers, P., Sidorenko, J., Karlsson Linnér, R., Fontana, M. A., Kundu, T., Lee, C., Li, H., Li, R., Royer, R., Timshel, P. N., Walters, R. K., Willoughby, E. A., … Cesarini, D. (2018). Gene discovery and polygenic prediction from a genome-wide association study of educational attainment in 1.1 million individuals. *Nature Genetics, 50*(8), 1112–1121. https://doi.org/10.1038/s41588-018-0147-3

Lodato, M., et al. (2015). Somatic mutation in single human neurons tracks developmental and transcriptional history. *Science, 350*, 94–98.

LoPresti, P. (2020). HDAC6 in diseases of cognition and of neurons. *Cells, 10*(1), 12. https://doi.org/10.3390/cells10010012

Lunden, J. W., Durens, M., Nestor, J., Niescier, R. F., Herold, K., Brandenburg, C., … Nestor, M. W. (2020). Development of a 3-D organoid system using human induced pluripotent stem cells to model idiopathic autism. In *Advances in neurobiology neurodevelopmental disorders* (pp. 259–297). https://doi.org/10.1007/978-3-030-45493-7_10

Maier, L. J., Ferris, J. A., & Winstock, A. R. (2018). Pharmacological cognitive enhancement among non-ADHD individuals-a cross-sectional study in 15 countries. *International Journal of Drug Policy, 58*, 104–112.

Marshall, P., & Bredy, T. W. (2016). Cognitive neuroepigenetics: The next evolution in our understanding of the molecular mechanisms underlying learning and memory? *NPJ Science of Learning, 1*(1). https://doi.org/10.1038/npjscilearn.2016.14

Mlecnik, B., Galon, J., & Bindea, G. (2018). Comprehensive functional analysis of large lists of genes and proteins. *Journal of Proteomics, 171*, 2–10. https://doi.org/10.1016/j.jprot.2017.03.016

Moran, D., Cohen, J. D., & Cohen, J. (2012). *Husserl dictionary*. Bloomsbury Academic.

Nestor, M. W., & Wilson, R. L. (2018). Can the use of CRISPR in humans result in decreased social justice for future stakeholders? *Ethics in Biology, Engineering and Medicine: An International Journal, 9*(1), 5–16. https://doi.org/10.1615/ethicsbiologyengmed.2019030551

Palma-Gudiel, H., & Fañanás, L. (2021). Prenatal exposures and behavioral epigenetics in human infants and children. In *Developmental Human Behavioral Epigenetics* (pp. 83–90). https://doi.org/10.1016/b978-0-12-819262-7.00004-0

Seay, G., & Nuccetelli, S. (2017). *Engaging bioethics: An introduction with case studies*. Routledge/Taylor & Francis Group.

Sinhababu, N. (2013). The desire-belief account of intention explains everything. *Noûs, 47*(4), 680–696. https://doi.org/10.1111/j.1468-0068.2012.00864.x

Tibbo, A. J., & Baillie, G. S. (2020). Phosphodiesterase 4B: Master regulator of brain signaling. *Cells, 9*(5), 1254. https://doi.org/10.3390/cells9051254

Toyoshima, F., Barton, A., & Grenier, O. (2020). Foundations for an ontology of belief, desire and intention. *Frontiers in Artificial Intelligence and Applications Formal Ontology in Information Systems*. https://doi.org/10.3233/faia200667

Wallace, K. (2000). Agency, personhood, and identity: Carol Rovan's the bounds of agency. *Metaphilosophy, 31*, 311–322. https://doi.org/10.1111/1467-9973.00151

Chapter 5
The Use and Access to CRISPR in Historically Socioeconomically Disadvantaged and Marginalized Communities

5.1 Introduction

We have discussed in earlier chapters the promise and practical application of CRISPR technology to treat and potentially ameliorate human genetic conditions. CRISPR technology is powerful because it can be used to directly and specifically target sequences of DNA in cells (Doudna & Charpentier, 2014). It allows biomedical researchers to understand the functional relationships between genes and proteins, and ultimately proteins and the function of cells (Nestor et al., 2014; Nestor et al., 2016). Somatic CRISPR editing represents the vast majority of work done using this technology. Until recently, it has been suggested that CRISPR germline editing was not possible without a significant amount of difficulty. This was supported by research using CRISPR in embryos that were not viable (Liang et al., 2015). Recent developments made famous by the "CRISPR-babies" controversy (Normile, 2018) have demonstrated that germline CRISPR editing should be taken seriously by the research community and is now being treated as potentially inevitable (Hurlbut, 2019).

In 2015, a team of scientists at Yat-sen University in Guangzhou, China, published the first germline human embryo editing using CRISPR (Liang et al., 2015). This work was quickly followed by a second Chinese group editing HIV-related genes in human embryos (Kang et al., 2016). After the publication of this research, CRISPR editing of human embryos in Sweden and the United Kingdom was approved by regulatory agencies (Callaway, 2016). In conjunction, Shoukhrat Mitalipov's group at Oregon Health & Science University became the first lab in the United States to CRISPR-edit human embryos to correct for a mutation in the *MYBPC3* gene (Ma et al., 2017).

The speed at which CRISPR has evolved from a basic science laboratory research tool to one used to manipulate the genetic composition of human embryos is astounding. A Chinese team already has CRISPR-genetically modified monkey

Adapted with permission from Nestor and Wilson (2018).

M. W. Nestor, R. L. Wilson, *Anticipatory Ethics and The Use of CRISPR in Humans*, https://doi.org/10.1007/978-3-030-98368-0_5

embryos, inserted them into surrogate mothers, and produced modified offspring (Niu et al., 2014). Other groups are approaching these techniques in humans (Midic et al., 2017). CRISPR-mediated embryonic editing represents a lucrative and attractive area for scientific endeavors with a significant potential impact on human civilization. While developing sets of guidelines, policymakers have not fully kept up with the rapid development of these technologies. We will discuss policy work done in this area and potential solutions in later chapters. Countries have chosen to ban embryonic modification while not fully considering that some governments may feel pressure from the scientific community to continue pursuing CRISPR-embryonic editing without developing "red lines" and treaties that legally bind countries to responsible biomedical research practices. Indeed, there is some support for germline CRISPR editing outside of medically necessary applications, and ethicists like Julian Savulescu of Oxford University have argued:

> There's huge variation between people in terms of fundamental characteristics and dispositions. When the science of genetics allows us to choose between the range of children that we could have, between those that will have better lives for themselves and be better functioning members of society, we ought to select those embryos rather than just tossing a coin. (Webber, 2016)

The argument made by Savulescu and others is that non-medical genetic enhancement increases the freedom of future generations because the characteristics of those generations would not be limited and would aid in allowing future generations to take advantage of a future in which gene editing exists (Glover, 2006). Thus, here we will focus only on the stakeholder perspective of future persons and what CRISPR technology will mean for them in the context of freedom and justice.

5.2 Considerations of Future Stakeholders

Savulescu's argument could be one possible logical conclusion based on mild ethical consequentialism from the stakeholder perspective of future individuals. Future persons may have several baseline intentions that stem from a consequentialist perspective. For example, future persons may intend a disease-free state equating to maximal healthiness. Future persons may also intend to live the happiest life possible, which entails the least amount of suffering. Future persons may also intend to live in a society where scientific progress is exercised to its maxima. Indeed, the intentions of both maximal health and maximal happiness may rely on a robust scientific enterprise to bring about technological improvements that enhance life. In this formulation, CRISPR represents a technological advancement that brings about the highest level of health by eradicating disease states that increase suffering and detract from the happiest life. So far, to future persons, CRISPR represents a technology that also increases their freedom by reducing or eliminating their reliance on medical interventions. This increase in freedom is conjoined with a future person's happiness. This moral argument for the advancement of CRISPR human germline editing can be buttressed from the perspective of Epicurean ethics:

> The necessary desires are for health of body and peace of mind; if these are satisfied, that is enough for a happy life. (Kauffmann & Baird, 2003)

In this context, from the perspective of future persons, it is our current moral imperative to pursue CRISPR editing of human embryos. We reached this conclusion in the earlier chapter on phenomenology and CRISPR concerning a view from the short-time horizon. Future persons are stakeholders that rely upon decisions we make about the scientific enterprise in the present. Therefore, they will see as moral those scientific decisions that maximize their happiness and freedom. Thus, our present scientific and policy-making actions concerning the use of CRISPR in humans will result in positive outcomes and happiness, while attempting to avoid negative outcomes and unhappiness from the perspective of future stakeholders.

5.3 Pluralist Universalism, Intuition Ethics, and CRISPR

Suppose we adopt a more comprehensive moral view from a potential future stakeholder perspective. In that case, we can see that the argument from happiness is partially undermined by the argument for *pluralist universalism* put forth by Robert Audi in his book "Moral Value and Human Diversity." Pluralist universalism consists of a combinatorial analysis of three main ethical philosophies and the work of virtue theorists. Audi chooses happiness from utilitarianism, justice from Kantian ethics, and freedom from Kantianism, utilitarianism, and Nozickian libertarianism. Audi develops pluralist universalism around the premise that happiness must be optimized without imposing on freedom or resulting in injustice, a lack of either of which will result in a lack of moral virtue. As such, Audi argues:

> Considerations of justice and freedom take priority (at least normally) over considerations of happiness; justice and freedom do not conflict because justice requires the highest level of freedom possible within the limits of peaceful coexistence, and this is as much freedom as any reasonable ideal of liberty demands. (Audi, 2007)

Therefore, we have a scientific contractual agreement with future stakeholders in the form of a social contract to maximize justice and freedom between current scientists and individuals advocating for CRISPR in humans. This view of CRISPR in the context of future stakeholders hinges on maximizing social justice as argued by Rawls:

> According to Rawls – to ascertain the principles of justice we should imagine ourselves as deciding upon fundamental ground rules for society: we must consider and decide upon the fundamental ground rules for society; but we must consider and decide upon these principles in a way that prevents us from showing any partiality to ourselves or to our family or friends. (Waymack & Hoppe, 2011).

By choosing CRISPR germline editing, we are showing partiality to our present selves without considering the lived experience of future persons (the importance we discussed in Chap. 3) and concerning social justice.

Going back to eliminating genes that cause depression via the use of CRISPR is illustrative. From a utilitarianist moral perspective, it seems reasonable to remove genes from human embryos that will result in depression. This, based on analysis by Layard et al., (2013) and Lyubomirsky et al., (2005), would result in lower levels of suffering and increased happiness.

To apply Audi's pluralist universalism to this social contract with future persons, we first must understand that this moral program is conducted from a common-sense *intuition ethics*-based point of view. Intuition ethics posits that future stakeholders have different mental models that underlie their common-sense moral intuitions, intermediate principals, and theoretic moral principles (Werhane, 2002). Paramount to how future persons frame moral decisions emerges from the intersection of these moral thought methods, but ultimately rests on common-sense moral intuitions. As we illustrated in Chap. 3, future moral decisions and perspectives are also shaped by the phenomenological world view (comprising the life-world and intersubjective experiences) of the stakeholders in the present.

Using a present mental model of intuition ethics, combined with Audi's pluralist universalism that places freedom and justice above happiness, we can create a moral imperative that respects the social contract with future stakeholders. For example, in the case of depression, we can pose the question, "should CRISPR be used to eradicate depression?" Although future persons might argue that doing so increases happiness and therefore is ethical, there is also a competing perspective based on our arguments made in Chap. 3 on the lived experience of future persons and pluralist universalism that using CRISPR to eradicate depression is not moral because it introduces harm for both current and future persons.

Current stakeholders may see the use of CRISPR as a restriction of personal freedom. This is compounded by the fact that current stakeholders may be pressured to reduce their current freedom when people choose to produce humans that only fit a "genetically healthy" category. This pressure to conform to a world that has been awash in gene-editing techniques is often referred to as an "instrumentalist" position, meaning that individuals are treated as instruments in this framework (De Wert et al., 2018). In this type of instrumentalism, CRISPR editing would result in decisions to use CRISPR that decrease future stakeholders' freedom and justice. Current stakeholders break the social contract with future stakeholders and reduce freedom and justice by using CRISPR by eliminating genetic diversity and narrowing the range of human experiences. This would potentially remove the freedom to experience an open future and undermine human dignity (Feinberg, 1980; Habermas, 2003; De Wert et al., 2018).

For instance, it has been shown that artists and writers generally exhibit higher levels of depression as compared to the general population (Jamison, 1989). Reducing rates of depression using CRISPR potentially restricts the lived experience and reduces the diversity of human condition that enhances both current and future stakeholder's lives. We can hypothesize that there will be unintended consequences of correcting genes that cause depression. Assuming that the problem of pleiotropy has been addressed for multigenic conditions such as depression, resolving depression in the general population produces an apparent ethical contradiction. On the one hand, CRISPR correction of genes that underlie depression increases happiness. Yet, on the other, they may deprive the world of the diversity of artists and writers that create works that contribute to happiness. This central problem emerges from an anticipatory analysis of CRISPR concerning certain types of genetic conditions that affect the mind, and one we partially addressed in Chap. 4.

This significant decrease in a broader range of experiences may significantly reduce the freedom and autonomy of future persons. Additionally, future stakeholder freedom with respect to reproduction will be limited to only creating progeny without such conditions. This may seem like a positive outcome, and it would be but with some limitations. Building a future with no progeny with depression may be one in which the richness of future lived experience is narrowed, using the example of artists and writers stated earlier.

From an intuition ethics standpoint, and in this context, future persons may have a negative ethical assessment of CRISPR, which is linked to more intermediate principles such as minimizing harm. In the case of future stakeholders, when both genetic diversity and the diversity of the lived experience are decreased, harm is induced and a reduction in freedom and justice, resulting in a broken social contract and an assessment that the use of CRISPR is immoral.

5.4 Social Justice and the Tyranny of Future Persons

Suppose we assume that future stakeholders were only utilitarians and Epicureans and overlay their imagined ethical perspective on the inevitability that CRISPR technology will advance at a rapid and relatively uncontrolled pace. In that case, we must understand how these stakeholder perspectives will act on current persons to decrease freedom and social justice.

As argued earlier, CRISPR can be classed as a disruptive and paradigm-shifting technology (Rogers, 2001). We can use another paradigm-changing technology as an analog for CRISPR: computers. At the advent of computing, it was argued that this technology would normalize socioeconomic divides and break down class and race boundaries. For this reason, we can imagine future stakeholders arguing that CRISPR technology should be used in humans. Not only will scientists be able to eradicate disease, but they might be able to enhance the abilities of disadvantaged groups and bring into harmonization of classes of individuals through mechanisms such as cognitive enhancement. Indeed, this argument has been made for the case of genetic modification in general, such that gene editing increases social justice by normalizing characteristics and traits that are naturally unevenly distributed through the population (Buchanan et al., 2000). This, in turn, increases "the equality of opportunity for every person." (De Wert et al., 2018).

The moral argument for using CRISPR in humans in this context relies on an inherently *ceteris paribus* perspective and assumes that access to this technology will be *uniform* and *ubiquitous*. We can look at the worldwide rollout of computing. Similar assumptions about access to computing technology were made, especially to historically disadvantaged socioeconomic groups. Computing access has neither been uniform nor ubiquitous, leaving current stakeholders in the digital divide (van Dijk & Hacker, 2003; Sakamura & Koshizuka, 2005). Countries and societies on the top of the socioeconomic pyramid have the most access to computing technology. This access, and the services around it, are skewed towards those individuals with the most resources within those societies.

The scientific enterprise in global societies primarily advances current arguments for using CRISPR in humans. These individuals tend to be elites on the upper end of the socioeconomic spectrum. As Daniel Sarewitz writes,

> Scientists are not elected. They cannot represent the cultural values, politics and interests of citizens — not least because their values may differ significantly from those of people in other walks of life. (Sarewitz, 2015)

For this reason, it is easy for scientists to posit ceteris paribus cases for the moral acceptance of CRISPR use in humans. These cases are presented to policymakers as expert opinion and drive current stakeholder ethical perspectives with little regard to the social contract with future persons, particularly those from disadvantaged groups. Sarewitz writes:

> There is no way to capture the full complexity of these issues from a scientific perspective. When new technologies are introduced into complex socio-technical systems, everyone is ill-informed about the risks. (Sarewitz, 2015)

This has been the case in underestimating the ability for historically underrepresented groups to have uniform and ubiquitous access to computing technology. This same problem exists for access to CRISPR technology and its potential advantages. Additional considerations contribute to disparate outcomes in historically disadvantaged communities concerning the use of CRISPR—these bottlenecks in accessing stem from historical inadequacies in biomedical research. Hildebrandt and Marron (2018) outline these as:

- Underrepresentation in clinical studies
- Disparate access to research benefits
- Mistrust of research

These bottlenecks represent disconnects between the research community and the greater public they serve. These barriers represent a real-world lived experience of a large group of present and future stakeholders. To this end, the Nuffield Council on Bioethics Working Party on Human Genome Editing and Human Reproduction recommended a more considerable public debate on genome editing (Nuffield, 2018). In the report, the Nuffield Council reinforced the principle that the ethics of germline gene editing rests on whether there are ethical frameworks that protect "the welfare of the future person and the principle of social justice and solidarity." In earlier chapters, we have already established that germline CRISPR editing, particularly of genes that affect memory and cognition, violates the welfare of future persons. In this chapter, we have argued that this type of genome editing potentially violates the principle of social justice.

As structured right now within the global discourse, CRISPR in humans has a high propensity for reducing social justice and freedom. It does this by limiting the lived experience of psychological and biological biodiversity and conferring advantages onto those already socioeconomically advantaged concerning access to CRISPR. By reducing justice, freedom, and happiness, the use of CRISPR in humans in the present does not consider the widest possible range of stakeholder perspectives. It breaks the social contract with future persons. Suppose we assume inequality in access to this technology. In that case, we can see clearly that the use of CRISPR in humans is inherently unethical from the broadest perspective of future stakeholders.

Our emphasis on assessing the ethical viability of CRISPR use in humans from an intuitive ethics and pluralist universalism perspective while acknowledging its scientific inevitability is the groundwork for starting an international discussion. Although this is beginning to happen, scientists and subject experts dominate most of this discussion. It has not been fully opened to exploring the viewpoints of all the stakeholders involved. As a result, the ethical analysis is distorted to reflect only perspectives that suggest minimal societal risks. This risk analysis does not consider the complexity of unintended consequences and other essential considerations that set the parameters for social justice and the global public good. Sarewitz codifies this idea:

> As new areas of contentious technology emerge, the way out of this situation is to let democratic deliberation lead the way in determining which values and world views ought to be protected and which sacrificed. (Sarewitz, 2015)

An example of how this works in other fields is NASA's Expert and Citizen Assessment of Science and Technology group. This group has led to the inclusion of non-expert public opinion in scientific and ethical discussions. A recent position paper by several genetics experts has begun to develop this dialogue, arguing for transnational cooperation (Pei et al., 2017). Another contemporary approach to this problem is the global observatory for gene editing, which aims to bring together stakeholders from non-scientific and scientific perspectives (Jasanoff & Hurlbut, 2018). These attempts to deal with the problem of germline CRISPR editing represent an overall approach put forth by the scientific community. This participation from scientists and elites is laudatory but does not move forward enough into capturing the social justice considerations of present and future stakeholders from disadvantaged communities. Inclusive debates and guideline development considering public opinion are fundamental to including the widest set of stakeholder perspectives in an anticipatory ethical analysis. Although important:

> In aggregate, public opinion can reflect cultural biases and undervalue the perspectives of marginalized groups. Indeed, relying too heavily on public opinion can disregard essential human rights considerations. (Halpern et al., 2019)

To this end, Halpern et al. (2019) propose the development of a Human Rights Impact Assessment that can be incorporated with public opinion and stakeholder analysis to assess the impact of gene editing techniques on underrepresented and marginalized communities. However, to codify change, it will be necessary for governments to set and act upon national policies. This type of engagement, only made possible through policymakers, is the right step to shape our future with CRISPR in humans. By having a global conversation open to non-scientists, we will increase social justice perspectives, enhance freedom and happiness, and make ethical choices for future stakeholders.

5.5 Policy Perspectives That Include Historically Marginalized Stakeholders

From a policy perspective, social justice considerations for those not members of the scientific/intellectual elite and who are members of traditionally marginalized communities extend beyond ethics and human rights considerations, as central as these are. Marginalized communities also do not have full access and participation (e.g., "a seat at the table") when it comes to other considerations for using CRISPR in humans. For example, CRISPR will change the geopolitical power balance in some aspects of biomedical research and provide opportunities for financial gain via commercialization biased toward developed countries. In addition, the role of CRISPR in shaping public perception of promised forthcoming treatments and its role in altering overall culture concerning the scientific enterprise will not be uniform or ubiquitous (Asquer & Krachkovskaya, 2021). Even within mature biomedical ecosystems such as the United States, there are varying levels of trust amongst different stakeholders, particularly those from historically disadvantaged communities (Baffoe-Bonnie, 2018). Therefore, any policy that develops a binding agreement (as we will argue for in Chap. 9) needs, as its central consideration, benefit sharing and distributive justice concerning the use of CRISPR in humans. From the perspective of healthcare providers, justice as a moral concept is concerned with determining the right way to treat people based upon concepts such as human rights and the notion of fairness. Distributive justice, in particular, is concerned with what is owed to people and what is due to them (Seay & Nuccetelli, 2017). An essential part of justice in health care is how healthcare resources are distributed. Human rights and the notion of fairness play an important role in allocating healthcare resources. Still, distributive justice plays a central role in adjudicating between different parties who make claims upon healthcare resources based upon health care needs:

> Distributive justice is the part of social justice concerned with what is fair in the allocation of benefits and burdens in society between individuals or groups. (Seay & Nuccetelli, 2017)

In this way, distributive justice (DJ) lays the foundation for benefit-sharing, which flows from a DJ framework. Baffoe-Bonnie (2018) reveals the vital connection between these two concepts. A diverse and comprehensive set of stakeholder perspectives from marginalized communities are a critical foundation for the development of any policy, binding agreement, or international treaty developed to address the use of germline CRISPR editing in humans:

> Benefit-sharing is connected to the principle of distributive justice and ensures that there is an ethical and fair distribution of new biotechnologies. In the field of genomic biotechnology, the Human Genome Organization committee on Ethics, Law, and Society has defined benefits as 'good(s) that contribute to the well-being of an individual and/or a given community' which 'transcend avoidance of harm (non-maleficence) in so far as they promote the welfare of an individual and/or of a community'. Benefit-sharing requires that benefits from research impact not only those who participate in research. (Baffoe-Bonnie, 2018)

Notably, the ideas of DJ and benefit-sharing for CRISPR policy and involving historically disadvantaged communities must not be thrust on to these communities from a perspective of techno-scientific colonialist paternalism, which can actively increase inequality concerning DJ and benefit-sharing by the mere fact that this perspective "…engenders epistemic injustice." (Arguedas-Ramírez, 2020). In times of geopolitical rivalry, Arguedas-Ramírez argues,

> ...the precautionary principle and other ethical standards of research are perceived as an obstacle in the techno-scientific race. (Arguedas-Ramírez, 2020)

These types of policy structures and debates are dominated historically by scientists and other elites in the technopolitical power centers in the Global North without full consideration from stakeholders in the Global South. This arrangement can be seen as an analog for potential inequalities that could be codified by the misrepresentation of marginalized communities during the development of binding agreements concerning germline CRISPR editing in humans. The idea of benefit-sharing itself can be distorted by techno-scientific colonialist paternalism in that this type of colonialist paternalism can dominate how the scientific community labels what a benefit is:

> I understand that human suffering is caused by deadly genetic disorders. However, it is necessary to beware of the dangers of therapeutic misconception, sometimes promoted by scientists and clinical researchers. We must also think about the risks for global society if this technology could be used, without controls, for other purposes beyond preventing Mendelian diseases. It is also important to consider that technological optimists will likely tend to accept therapeutic misconceptions far more easily than technological pessimists would do. (Arguedas-Ramírez, 2020)

Therapeutic conceptions, according to technological optimists, are primarily grounded in a northern hemisphere, western-oriented perspective, and this perspective may not be shared globally by other cultures, governments, experts, or laypeople. It is vitally important to not only engage in a deep conversation about conceptions about CRISPR-mediated therapeutics with these communities but to do so rigorously and quantitatively, such that the viewpoints of these diverse sets of stakeholders are adequately addressed and represented in policies that drive an anticipatory governance structure around the use of CRISPR for germline editing in humans.

References

Arguedas-Ramírez, G. (2020). Ethics and global governance of human germline genome editing: The problem of techno-scientific colonialist paternalism. *The CRISPR Journal, 3*(2), 83–88. https://doi.org/10.1089/crispr.2019.0045

Asquer, A., & Krachkovskaya, I. (2021). A bibliometric analysis of research on CRISPR in social sciences and humanities. *Re:GEN Open, 1*(1), 14–23. https://doi.org/10.1089/regen.2021.0007

Audi, R. (2007). *Moral value and human diversity*. Oxford University Press.

Baffoe-Bonnie, M. S. (2018). A justice-based argument for including sickle cell disease in CRISPR/Cas9 clinical research. *Bioethics, 33*(6), 661–668. https://doi.org/10.1111/bioe.12589

Buchanan, A., Brock, D. W., Daniels, N., & Wikler, D. (2000). *From chance to choice. Genetics & Justice*. Cambridge University Press.

Callaway, E. (2016, April 19). *Gene-editing research in human embryos gains momentum*. Retrieved March 22, 2018, from https://www.nature.com/news/gene-editing-research-in-human-embryos-gains-momentum-1.19767

De Wert, G., Heindryckx, B., Pennings, G., Clarke, A., Eichenlaub-Ritter, U., Van El, C. G., Cornel, M. C. (2018). Responsible innovation in human germline gene editing: Background document to the recommendations of ESHG and eshre. *European Journal of Human Genetics, 26*(4), 450-470. doi:10.1038/s41431-017-0077-z.

Doudna, J. A., & Charpentier, E. (2014). Genome editing. The new frontier of genome engineering with CRISPR-Cas9. *Science, 346*(6213), 1258096.

Feinberg, J. (1980). The child's right to an open future. In W. Aiken & H. LaFollette (Eds.), *Whose child? Children's rights, parental authority and state power* (pp. 124–153). Rowman & Littlefield.

Glover, J. (2006). *Choosing children. The ethical dilemmas of genetic intervention*. Clarendon Press.

Habermas, J. (2003). *The future of human nature*. Polity Press.

Halpern, J., Ohara, S. E., Doxzen, K. W., Witkowsky, L. B., & Owen, A. L. (2019). Societal and ethical impacts of germline genome editing: How can we secure human rights? *The CRISPR Journal, 2*(5), 293–298. https://doi.org/10.1089/crispr.2019.0042

Hildebrandt, C. C., & Marron, J. M. (2018). Justice in CRISPR/Cas9 research and clinical applications. *AMA Journal of Ethics, 20*(9). https://doi.org/10.1001/amajethics.2018.826

Hurlbut, J. B. (2019). Human genome editing: Ask whether, not how. *Nature, 565*(7738), 135–135. https://doi.org/10.1038/d41586-018-07881-1

Jamison, K. R. (1989). Mood disorders and patterns of creativity in British writers and artists. *Psychiatry, 52*(2), 125–134.

Jasanoff, S., & Hurlbut, J. B. (2018). A global observatory for gene editing. *Nature, 555*(7697), 435–437. https://doi.org/10.1038/d41586-018-03270-w

Kang, X., He, W., Huang, Y., Yu, Q., Chen, Y., Gao, X., Sun, X., & Fan, Y. (2016). Introducing precise genetic modifications into human 3PN embryos by CRISPR/Cas-mediated genome editing. *Journal of Assisted Reproduction and Genetics, 33*(5), 581–588. https://doi.org/10.1007/s10815-016-0710-8

Kauffmann, W., & Baird, F. E. (2003). *Philosophic classics. Ancient philosophy*. Prentice Hall.

Layard, R., Chisholm, D., Patel, V., & Saxena, S. (2013). Mental illness and unhappiness. SOEP papers on Multidisciplinary Panel Data Research, No. 600.

Liang, P., Xu, Y., Zhang, X., Ding, C., Huang, R., Zhang, Z., Lv, J., Xie, X., Chen, Y., Li, Y., Sun, Y., Bai, Y., Songyang, Z., Ma, W., Zhou, C., & Huang, J. (2015). CRISPR/Cas9-mediated gene editing in human tripronuclear zygotes. *Protein & Cell, 6*(5), 363–372. https://doi.org/10.1007/s13238-015-0153-5

Lyubomirsky, S., Sheldon, K. M., & Schkade, D. (2005). Pursuing happiness: The architecture of sustainable change. *Review of General Psychology, 9*(2), 111–131.

Ma, H., Marti-Gutierrez, N., Park, S. W., Wu, J., Lee, Y., Suzuki, K., Koski, A., Ji, D., Hayama, T., Ahmed, R., Darby, H., Van Dyken, C., Li, Y., Kang, E., Park, A. R., Kim, D., Kim, S. T., Gong, J., Gu, Y., … Mitalipov, S. (2017). Correction of a pathogenic gene mutation in human embryos. *Nature, 548*(7668), 413–419. https://doi.org/10.1038/nature23305

Midic, U., Hung, P. H., Vincent, K. A., Goheen, B., Schupp, P. G., Chen, D. D., Bauer, D. E., VandeVoort, C. A., & Latham, K. E. (2017). Quantitative assessment of timing, efficiency, specificity and genetic mosaicism of CRISPR/Cas9-mediated gene editing of hemoglobin beta gene in rhesus monkey embryos. *Human Molecular Genetics, 26*(14), 2678–2689. https://doi.org/10.1093/hmg/ddx154

Nestor, M. W., Artimovich, E., & Wilson, R. L. (2014). The ethics of gene editing technologies in human stem cells, *5*(4), 323–338. https://doi.org/10.1615/EthicsBiologyEngMed.2015013732

Nestor, M. W., Phillips, A. W., Artimovich, E., Nestor, J. E., Hussman, J. P., & Blatt, G. J. (2016). Human inducible pluripotent stem cells and autism spectrum disorder: Emerging technologies. *Autism Research, 9*(5), 513–535. https://doi.org/10.1002/aur.1570

Nestor, M. W., & Wilson, R. L. (2018). Can the use of CRISPR in humans result in decreased social justice for future stakeholders? *Ethics in Biology, Engineering and Medicine: An International Journal, 9*(1), 5–16. https://doi.org/10.1615/ethicsbiologyengmed.2019030551

Niu, Y., Shen, B., Cui, Y., Chen, Y., Wang, J., Wang, L., Kang, Y., Zhao, X., Si, W., Li, W., Xiang, A. P., Zhou, J., Guo, X., Bi, Y., Si, C., Hu, B., Dong, G., Wang, H., Zhou, Z., … Sha, J. (2014). Generation of gene-modified cynomolgus monkey via Cas9/RNA-mediated gene targeting in one-cell embryos. *Cell, 156*(4), 836–843. https://doi.org/10.1016/j.cell.2014.01.027

Normile, D. (2018). Shock greets claim of CRISPR-edited babies. *Science, 362*(6418), 978–979. https://doi.org/10.1126/science.362.6418.978

Nuffield Council on Bioethics. (2018). *Genome editing and human reproduction: Social and ethical issues*. Nuffield Council on Bioethics.

Pei, D., Beier, D. W., Levy-Lahad, E., Marchant, G., Rossant, J., Izpisua Belmonte, J. C., Lovell-Badge, R., Jaenisch, R., Charo, A., & Baltimore, D. (2017). Human embryo editing: Opportunities and importance of transnational cooperation. *Cell Stem Cell, 21*(4), 423–426. https://doi.org/10.1016/j.stem.2017.09.010

Rogers, E. M. (2001). The digital divide. *Convergence, 7*(4), 96–111. https://doi.org/10.1177/135485650100700406

Sakamura, K., & Koshizuka, N. (2005). Ubiquitous computing technologies for ubiquitous learning. In *IEEE international workshop on wireless and mobile technologies in education (WMTE'05)* (pp. 11–20). https://doi.org/10.1109/WMTE.2005.67

Sarewitz, D. (2015, June 23). *CRISPR: Science can't solve it*. Retrieved March 22, 2018, from https://www.nature.com/news/crispr-science-can-t-solve-it-1.17806

Seay, G., & Nuccetelli, S. (2017). *Engaging bioethics: An introduction with case studies*. Routledge/Taylor & Francis Group.

van Dijk, J., & Hacker, K. (2003). The digital divide as a complex and dynamic phenomenon. *The Information Society, 19*(4), 315–326. https://doi.org/10.1080/01972240309487

Waymack, M. H. (2011). Ethical theories part 1. In E. A. Hoppe (Ed.), *Ethical issues in aviation*. Ashgate Press.

Webber, D. (2016, January 2). *As a species, we have a moral obligation to enhance ourselves*. Retrieved March 22, 2018, from https://ideas.ted.com/the-ethics-of-genetically-enhanced-monkey-slaves/

Werhane, P. H. (2002). Moral imagination and systems thinking. *Journal of Business Ethics, 38*(1–2), 33–42.

Chapter 6
Future Use of CRISPR: Gene Drive

6.1 Introduction

CRISPR can be used to edit the germline of embryos, implying that any corrected or introduced mutations can be subsequently passed on to offspring. Indeed, some of the main arguments dismissing any slippery-slope implications of the use of CRISPR in humans, and particularly the use of CRISPR to develop non-medical enhancements of traits such as IQ, is that these types of traits are not 100% heritable (De Wert, 2018). Additionally, the idea that using CRISPR to edit genes that contribute to any multigenic complex trait will have a negative effect on subsequent generations has been challenged as being scientifically impossible on these grounds (see Chap. 4). The limitations of Mendelian genetics prevent rapid changes in the populations of some CRISPR-edited organisms. By combining CRISPR with gene drive, which allows for "stimulating biased inheritance of particular genes to alter entire populations," edited genes can be spread throughout an offspring population at rates significantly higher than Mendelian genetics would predict (Esvelt et al., 2014). In rapidly reproducing populations like insects, the combination of these systems has the potential to produce a wholesale change of the genetic characteristics of entire populations rapidly (Esvelt, 2014). CRISPR and gene drive has been proposed to be used to induce negative ecological impacts on invasive insects or organisms and prevent the reproduction of malaria-containing mosquito populations (Bogardus & Saplakoglu, 2017). These uses imply that this technology can be easily controlled and benefit humans (Nestor & Wilson, 2018).

Although wholesale changes can occur in the population genetics of rapidly reproducing species, changes in organisms with a long lifespan would take decades or centuries using CRISPR and gene drive. This, combined with the notion that over long periods of time, natural selection may undo the effects of CRISPR and gene drive and artificial mate control can limit their scope, has also been used as an argument against the effectiveness of these technologies in humans. However, as with

Adapted with permission from Nestor and Wilson (2020)

M. W. Nestor, R. L. Wilson, *Anticipatory Ethics and The Use of CRISPR in Humans*, https://doi.org/10.1007/978-3-030-98368-0_6

every revolutionary scientific discovery and its relation to society, there is the risk of unintended uses or consequences for future persons. Here, we use an anticipatory ethical framework to explore the potentially negative effects for future persons concerning CRISPR combined with gene drive, a new and rapidly developing technology like CRISPR itself, to permanently alter the human genome for future stakeholders. We argue that the unintended consequences of using a combination of CRISPR and gene drive present a real risk to future persons, and thus, policy should be developed as soon as possible to address this anticipated risk to justice for future stakeholders.

Could CRISPR and gene drive technologies affect human populations? We believe that anticipatory ethics can be used to attempt to address the ethical issues raised by asking this question. As we reviewed in Chap. 2, anticipatory ethics examines potential ethical problems with emerging technology while this technology is still in the early stages of development. Consequently, ethical issues need to be identified at the research and development stage of technological progress and explored as possibilities experienced by potential stakeholders. The most substantial clinical work has been carried out via somatic editing using CRISPR, as stated in earlier chapters. Although this method of gene editing in cells is, for most, not controversial (in our analysis), somatic CRISPR editing does introduce several ethical dilemmas that are beyond the scope of this manuscript (Polcz & Lewis, 2016). The progress in germline CRISPR editing is not as advanced as somatic editing at the time of writing this manuscript.

Although anecdotally, most scientists are not in favor of germline CRISPR editing as currently practiced, there is no formal international regulatory system to keep this work in check beyond difficult to enforce recommendations, which we will review in Chap. 8. Both CRISPR and gene drives are technologies that are in the early phases of development and implementation. As such, they are ripe for an anticipatory ethical analysis and suggested courses of action for policymakers. CRISPR's ability to directly and efficiently edit the human genome with a high level of scalability allows it to be combined with gene drive technology.

Gene drive technology is a system whereby mutations or corrections to the organism's genetic code can be "driven" through subsequent populations (Esvelt et al., 2014). This occurs via a genetic engineering system in which offspring inherit mutations or corrections at rates much higher than those predicted by Mendelian genetics. In a typical Mendelian system, it is expected that a given gene will be transmitted to an offspring at a 50% probability. Some genes will decrease in frequency or expression over a population's developmental timeline, taking genetic fitness into account. In extreme cases, genes can disappear from the population. Stated simply, in the case of gene drives, CRISPR is used to introduce specific mutations and alterations into the host genome. After introducing these mutations and alterations, gene drive can be used to ensure that those mutations are passed to subsequent offspring at rates approaching one hundred percent inheritance. As such, gene drive technology, combined with the use of CRISPR, can make wholesale alterations in the genetic characteristics of entire populations of organisms.

The most striking example of CRISPR combined with gene drive is introducing a Zika-resistant gene into a native mosquito population along with the guide RNA and Cas protein that drive replication of this gene in offspring (the gene drive). In a Zika-infested population of mosquitoes, the gene for resistance would be passed on to all subsequent offspring with near one hundred percent transmission, almost guaranteeing that the vast majority of mosquitoes would be Zika-free within one generation (Gardner, 2017). Another positive application of this technology in mosquitoes is to confer malaria resistance in the wild. Using gene drive to spread a malaria-resistant system to wild populations of mosquitoes using CRISPR, the altered gene and gene drive system can alter entire mosquito species and play a significant role in improving world health. In the context of these potentially positive effects of gene drive systems, we set out to ask two major questions:

1. In what sense can CRISPR and gene drive affect human populations?
2. What would be the ethical implications if such a system were implemented in humans?

6.2 Applications of CRISPR "Gene Drive" Technology on Humans

Gene drive represents a way for genes that modulate traits to be pushed to offspring and perpetuate these genes through the species (Akcakaya et al., 2018). For example, the technology as it currently exists can be used to positively affect the human germline by decreasing or eliminating the number of mutations associated with genes involved in highly heritable monogenic conditions such as familial hypercholesterolemia achondroplasia, or sickle-cell anemia (Schwank et al., 2013). With continued technological development, the methods used will eventually be perfected and used for more complex diseases.

Scientists and researchers at the Wyss Institute at Harvard, where the technology is currently being perfected, do not think that gene drives can affect human populations without taking an extended period—perhaps centuries. This is partially due to the life cycle of organisms harboring the gene drive system. For example, mosquitoes have a very high population turnover rate (10 days for males) and are good candidates for gene drive systems to eradicate disease quickly across most organisms. In organisms with a longer lifespan, affected genes would take a long time to spread through populations:

> For example, if we introduced a trait into elephants (which have a long generation time like humans) using a gene drive today, there would only be four times as many elephants with that trait in 100 years as if we didn't use a drive.

One of the major proponents of gene drives, Kevin Esvelt, argues that because the gene drive genes would be so slow to spread in a population, there is time to create and release an "immunizing reversal drive" for the gene drive, thus rendering low risk in the immediate term. The overarching argument from this pro-human gene drive perspective is that this technology represents a minimal risk. As such, with slow reproduction cycles, alterations could be made to organisms that would improve life, which could be reversed if necessary via an immunizing reversal drive. Thus, gene drives should be viewed as a technology that should produce an overarching positive social outcome.

The arguments put forward by scientists and researchers in the pro–gene drive camp are problematic from the viewpoint of anticipatory ethics because these arguments make questionable assumptions:

1. As they exist today, CRISPR and gene drive technology will not continue to develop and improve. Take the example of CRISPR, which has already improved exponentially in its efficiency since its introduction only 5 years ago. As gene drives become more efficient and encompass more than one gene, it may be the case that reversal mechanisms become unwieldy or impossible to use.
2. There is an assumption that gene drives in organisms cannot be shielded from immunizing reversal drive. It is reasonable to conjecture that gene drives can be designed in such a way to prevent or diminish the effect of immunizing reversal drive. Even if an immunizing reversal drive was 80–90% effective, this leaves 10% of the population with a non-reversible gene drive and potentially creates special minority classes of organisms.
3. In the case where gene drive combined with CRISPR is used in humans, the pro–gene drive argument assumes that immunizing reversal drives will be given equal access to all populations in need of them. This presumes that the immunizing technology will be uniform and ubiquitous. This has not been the case for other large-scale technologies, such as the roll-out and worldwide access to computing technology (Gulati, 2008).
4. There is a risk of uncontrolled gene flow via crossbreeding in species that can crossbreed, allowing for infiltration beyond the initial target.
5. Uncontrolled mutant genes not initially introduced by CRISPR but occurring due to its use could spread into offspring via off-target effects of gene drive system, resulting in massive amplification of deleterious mutations. As it has been shown that CRISPR does have some off-target effects (Fu et al., 2013; Ma et al., 2017; Sun & Zhao, 2014), these off-target gene effects can potentially be passed to subsequent generations before being detected, harming future humans.
6. Because concerns in other areas of biology have not yet materialized (presumably due to the technology not being perfected or widespread), it is safe to assume the same for gene drive.

Thus, even if we grant the pro–gene drive arguments and the claim that human changes would take a long time to materialize, there is the potential for adverse outcomes, including those of long-term unintended consequences and off-target effects, which need to be considered by policymakers. How then do we assess long-term consequences and risks?

6.3 Anticipatory Analysis of Gene Drive in Humans

How should ethical issues be identified, and what principles should be employed to analyze these issues? To include the broadest range of stakeholder perspectives and to introduce social justice into the discussion of CRISPR and gene drives, we introduce a fivefold scheme. To gain relevance and understanding for each party and stakeholder potentially affected by this technology, the range of ethical concepts, principles, and systems used in this analysis includes principles that fall into these categories (Sun & Zhao, 2014).

1. There must be highly specific principles that apply to particular circumstances. This includes the principles used by ordinary people in the everyday world in ordinary circumstances. These principles could be as simple as positive and negative reactions or intuitions by ordinary people to what they perceive to be the effects of technology.
2. There are intermediate principles such as harm minimization, justice, freedom, and happiness, which are not highly specific but not acknowledged as normative or specific technical, ethical principles.
3. There are traditional theoretical or normative ethical principles such as consequentialism, deontology, and social contract theories.
4. There are ethical principles developed by professionals in an area of expertise (e.g., medicine) that the individuals within that field endorse. Principles in codes of ethics fall into this category.
5. There is the possibility of *emergent* ethical principles arising due to the uniqueness and rapid pace of development of new technologies. New technologies can trigger the need for new ethical principles.

We shall introduce 1–4, and then 5. A discussion can be conducted from a commonsense intuition ethics-based point of view to deal with the first level. Once intuitions are identified, ethical analysis can correlate commonsense moral intuitions to intermediate principles and theoretical moral principles. To assess CRISPR-mediated gene drives, this technology will be morally assessed based upon how it affects and impacts the stakeholders affected by it. Different stakeholders perceive and are affected by the technology differently. As discussed in earlier chapters, one can intuit moral perspectives based on stakeholder analyses where ordinary individuals during their lived experience have positive and negative reactions to gene-editing technology.

For example, in the case of a condition such as autism, one can anticipate two juxtaposed moral perspectives from the parent of an individual with autism and the individual when posed the question, "should gene editing and gene drive be used to eradicate autism?" The parent of an individual with autism might argue that it is moral to use gene drive to eradicate potential suffering. In contrast, an individual with autism may see it as an immoral deleting of their uniqueness and individual freedom to have their genes replaced by those expressed by neurotypical individuals. In this example, these positive and negative reactions can be linked to more intermediate principles such as minimizing harm (see category #2 above).

The concept of "harm" can be applied from an ordinary intuitive perspective (category #1) and the analysis can be conducted at that level. However, the analysis can be applied using category #1 and "harm" as defined by a professional expert position put forth by those in an area of expertise such as doctors (category #4). This illustrates the non-linearity and non-mutual exclusiveness of the categories we have outlined in our applied anticipatory ethical analysis (AEA).

To deepen this ethical analysis using the stakeholder perspective combined with category #3, let's consider a stakeholder analysis that employs normative ethical principles that correspond with ethical intuitions. This analysis will provide a foundation for conducting an ethical assessment of CRISPR and gene drive technologies. We assume a basic theory of action based upon an Aristotelian theory of action. A series of normative ethical principles will be selected as possible guides for formulating intentions, helping us decide how to act, and reflecting upon short- and long-term outcomes of our actions. These principles can be selected as foundational for an ethical analysis because they can be related to our common sense, ethical intuitions. We will employ justice, freedom, and happiness as the foundational principles for conducting our ethical analysis. *Honest Business* states that justice, freedom, and happiness can also be directly related to underlying substantial ethical principles (Cuilla et al., 2007). The principle of justice is related to a sense of fairness and contractarian rights and linked to actions that members of society perform within social circumstances. According to this view, members of a society have a series of moral contracts with one another and are required to perform actions.

Responsibility for actions can also be related to and to rest upon the intentions of agents and the choices for actions made based upon these intentions. A duty- or rule-based system claims that human actors have responsibilities to one another based upon autonomously endorsed duties and rules. Intentions need to be based upon duties and rules. Finally, actions and the consequences of our choices and actions, both short- and long-term, are taken to be the essence of moral reasoning and judgments. Actions and the consequences are the bases of possible ethical judgments and choices. The ultimate goal of these judgments with actions based upon them is to obtain a result that maximizes positive outcomes and happiness.

How do we reconcile the different conclusions to which each of these principles leads when we attempt to assess the use of CRISPR and gene drive technologies? Using the same principle developed by Robert Audi in *Moral Value and Human Diversity*, and described in Chap. 5, when ethical judgments based upon these principles conflict with one another,

> Considerations of justice and freedom take priority (at least normally) over considerations of happiness; justice and freedom do not conflict because justice requires the highest level of freedom possible within the limits of peaceful coexistence, and this is as much freedom as any reasonable ideal of liberty demands. (Audi, 2007)

Based upon this discussion, one starting point for our AEA is to apply categories 1–4 to various stakeholder perspectives referred to above and examine how those occupying these perspectives might perceive CRISPR and gene drives from the standpoint of *intuition ethics* overlaid onto our categorical system. We assume each of the stakeholders on our analysis (researchers/scientists, current ordinary persons, future ordinary persons) has intentions, performs actions, and produces outcomes with short- and long-term consequences. This style of analysis employs intuitions, intermediate principles, normative ethical principles, and principles from codes of ethics as first principles. Each of these principles is taken to be a correct and legitimate basis for making sound ethical judgments. The value of this type of analysis can be defended as an exploration of what each of these principles would say about an issue, such as, for example, CRISPR and gene drive technologies. To show the power of this technique as a calculus for anticipatory ethical considerations, we present an abbreviated analysis by examining just three stakeholder perspectives.

As stated in previous chapters, a complete stakeholder analysis of an anticipatory biomedical ethics problem is qualitatively and quantitatively complex. If done correctly could involve tens of thousands of stakeholders and equal numbers of perspectives. For an example of quantifying such a system, please see Babar et al. (2015).

Stakeholder #1: Researchers and Scientists
Intentions

Develop CRISPR and gene drive technology to create changes in organisms to alleviate disease vectors in the case of mosquitoes and genetic disorders in the case of humans. There is a duty to eliminate mosquitoes that carry deadly diseases for humans. These technologies are not harmful as they can theoretically be reversed. They will reduce suffering as defined by professional medical societies by decreasing rates of disease.

Actions

We need to refine and develop CRISPR and gene drives in mosquitoes and humans by increasing funding for research and development. This funding will increase basic research and clinical research and allow those in the field to assess the applications of the technology.

Consequences

If successful, the short-term consequences of using CRISPR and gene drive-in model organisms will include eradicating mosquito-borne diseases and the development of potential therapies for complex disorders. These results will possibly lead to the long-term consequences of eradicating and eliminating illnesses.

Stakeholder #2: Current Ordinary Persons
Intentions

Minimize current suffering and harm, and increase overall justice, freedom, and happiness.

Actions

Support funding for research programs that promote the use of CRISPR and gene drive for mosquitoes and humans in the hope that these technologies will lead to less disease and illness.

Consequences

If successful in the short term, the consequences for supporting research programs that develop technologies for the use of CRISPR and gene drive in mosquitoes and humans are the same for researchers and scientists.

Stakeholder #3: Future Ordinary Persons

Intentions

Living as healthy lives as possible while maximizing happiness, freedom, and justice.

Actions

Support funding for research programs that promote the use of CRISPR and gene drive for mosquitoes and humans in the hope that these technologies will lead to less disease and illness. Continuing to support research programs in the future at the behest of scientists/researchers who promise a long-term increase in happiness, even when short-term adverse effects may be evident.

Consequences

When those who exist in the future look back upon the development of these technologies, they may look at what we thought of as short-term and long-term consequences. They may ask what the intentions were of those who used CRISPR and gene drive technology in the past and whether they considered the long-term effect on future persons. They may look back upon the introduction of technologies that may have altered the diversity of ecosystems. This may have unintended deleterious consequences. The mosquito, for example, can be used as a vector to introduce disease directly or can itself be eliminated from the ecosystem, destroying crops or other food sources. This technology can be used to alter the human germline; elimination of genetic diversity would narrow the range of human differences. It may be the case that future persons would disagree with altering the human genome while arguing for eradicating the disease. A significant decrease in a wider range of experiences and differences may significantly reduce freedom and autonomy in both cases. Human freedom with regard to reproduction may be limited to creating progeny who does not have genetic disorders. Predecessors and the items they valued will have determined decisions related to happiness, justice, and freedom. Freedom will be limited when people will only produce humans who fit a predetermined "genetically healthy" category or profile. Because "... justice requires the highest level of freedom ..." (Audi, 2007), this decrease in justice implies that freedom has not been achieved.

A second style of ethical analysis can be deployed to conduct an AEA. This analysis begins from a different direction and was introduced as category #5 above. With developing technologies that are already in the process of being introduced or that are already introduced, a comparison of cases can lead to the discovery of emerging ethical issues. As stated above, CRISPR and gene drive technology have already been introduced in the context of mosquitoes. Comparing these cases and the stakeholders' intentions, actions, and short- and long-term consequences points to the claim that researchers, scientists, and policymakers have special responsibilities.

6.4 The Case of Emergent Ethical Principles Arising Due to the Uniqueness and Rapid Pace of Development of New Technologies

The problem with rapidly developing technologies is that they expose the limitations of ethical analysis according to 1–4 above. George Lucas offers a different strategy that can be applied to CRISPR and gene drive technologies and cases. Although used in a different context, Lucas makes the following statement:

> How do we handle uncertainty, including especially the uncertainty attached to novel developments, new technologies, and contrasting or competing ways of living—how do we grow comfortable with the new and the novel? How, in particular, do we discern the appropriate governing principles, or "rules of the game," when we don't know precisely what the "game" is, let alone its rules? (Allhoff et al., 2016)

Employing the ideas of Alister Macintyre and Aristotle, Lucas develops what he calls "the methodology of uncertainty." The method of Lucas works in the following way:

> Basically the method is this: when we are uncertain about what's going on, or what the rules or principles that properly apply to our actions are, we start by gathering all the relevant data we can about the practice in question—for example, moral customs in different cultures, various kinds of constitutions, and political arrangements (or, in our case, various activities taking place in the cyber domain). We begin intuitively to discern better from worse practices, based in large part upon their relative degree of "operational effectiveness." We also draw comparisons between what we know, and what we are newly confronting. We engage in reflection, dialectic, and sustained philosophical argument, in order to analyze both the practices in question and our comparative experiences with their results. We extrapolate from the known to the unknown, and gradually begin to build what Aristotle termed the Archai, the "first principles" governing practice, from our varied experiences of, and reactions to, the variant forms of the practices themselves. (Allhoff et al., 2016)

Lucas advocates a method of analysis that starts from a different direction than 1–4 referred to above. We begin by gathering relevant data about the practices in question, in this case, CRISPR and gene drives. What do we know about how these technologies have been practiced? With CRISPR and gene drives, this would mean examining the research into and applications of the technologies.

> Since early 2015, laboratory scientists have published four proofs-of-concept showing that a CRISPR/Cas9-based gene drive could spread a targeted gene through 100% of a population of yeast, fruit flies, or mosquitos. Researchers have been studying these natural mechanisms throughout the 20th century but, until the advent of CRISPR/Cas for gene editing, have not been able to develop a gene drive ... Most research to date is focused on controlling or altering organisms such as mosquitoes that transmit infectious diseases to humans. (National Academies of Science, 2016)

On the surface, we should end our AEA at the point where CRISPR is used in combination with gene drive in mosquitoes. We could posit, as others do, that:

> ... gene-editing experiments have been conducted thus far by a small community of scientists, with teams that include individuals with different types of expertise. In such teams only a couple of individuals may have the skills to insert the drive in the animal ... Without

> a clear and detailed understanding of the range of social and technical factors that cause scientists to succeed or fail in their gene-drive endeavors, threat estimates can only rely on speculation and fantasy rather than fact. (Ben Ouagrham-Gormley & Vogel, 2016)

However, the charge of an anticipatory ethical analysis is to develop possible outcomes that develop emergent ethical principles that outpace the rapid deployment of new technologies. Moreover, these analyses must make reasonable assumptions about the maturation and outcome of the technology in the absence of its full implementation in the present. Therefore, we must ask from a policy perspective:

- What are the moral customs for carrying out research on the populations in question?
- What are the customs for applications of methods that have never been applied before?
- What is known and what is unknown about what we are confronting?

From reflecting on these questions in the context of the categories outlined above and discussion about possible outcomes, we begin to build a policy framework for CRISPR and gene drive in humans.

6.5 Policy Considerations for the Use of Gene Drive

For policymakers, what are the factors that need to be considered, and what are the ethical and social issues? We can begin our analysis by asking, what are our intuitions about CRISPR and gene drives? Here we outline a set of considerations for policy-making surrounding CRISPR and gene drive:

- As CRISPR and gene drive technologies continue to develop, increased ease of use will occur.
- What are the risks of the technology? Who is determining the risk profile concerning the economic benefit from implementing this technology?
- A model of participatory decision-making is needed so that all stakeholders can understand what the risks are.
- Will reversal agents be equally and ubiquitously accessible to all segments of the population, regardless of social or economic status?
- What are the unintended consequences? Could a chain reaction of irreversible consequences be set off in off-target populations?
- There are issues of informed consent for the current generation. How can informed consent be obtained from the wide range of people who stand to be affected by the technologies?
- Issues of informed consent arise for and extend into future generations. How would it be possible to obtain informed consent from future generations affected by the technology?

- Further technological development will include an increase in the speed of technological change, and ethical and social analysis must attempt to keep pace with this change. Can a sound policy be informed this way?
- The regulatory environment must include discussion of the influence of these technologies upon the global public good and social justice based upon how CRISPR and gene drive technologies will become normalized concerning the lives of ordinary persons.

6.6 Anticipating Ethical Implications for the Future

Based on our intuitions, what ethical issues can we anticipate emerging from the development of CRISPR and gene drive technology? Given what has occurred, what are the practices we should strive for?

1. There is an essential distinction between remediation, modification, and enhancement that needs to be considered. Using CRISPR and gene drive to eliminate diseases and therapeutic purposes is different from modifying using these technologies for "enhancement" purposes.
2. The aims, intentions, and goals of those using the technology and the short- and long-term consequences must be considered. The proposed use of the technology must be based upon a sound understanding of the current state of the technology.
3. Technologies organize activities, alter relationships, and structure choices. It is essential to realize that CRISPR and gene drive technologies are not value-neutral. These technologies shape the conditions of human existence and human relationships, and they condition human interactions while simultaneously influencing our choices.
4. While it is important to see how this technology is being used now, an ethical analysis must avoid focusing exclusively on the use of technology in the current form; it must also consider what the future use of this technology may entail.
5. Those identifying, analyzing, and addressing issues raised by technologies such as CRISPR and gene drive need to embrace an interdisciplinary approach. Those involved in all aspects of research and development of the technology need to have a voice in deciding to use the technology. Multidisciplinary collaboration is crucial for stakeholders to communicate about the technology, understand the technology, and assess the technology's value.
6. It is crucial to understand in the context of this analysis that CRISPR/gene drive will continue to evolve and mature, leading to a more sophisticated technology. This may reduce some risks (such as off-target effects and immune response) and increase the therapeutic viability of this system. Indeed, it may be possible to engineer technologies that will reverse or stop gene drive, as Dr. Esvelt has suggested (we will discuss this in the next chapter). However, these improvements may introduce a new set of anticipatory ethical consequences. For example, if

gene drives are reversible and CRISPR has no off-target effects, is it ethical to eradicate all human diseases in favor of a genetically homogenous population? What does the eradication of manic depression, for example, mean to the lived experience of society when it has been demonstrated that there is a high correlation between manic depression and artistic creativity? (Andreasen, 2008)

Consequently, ethicists must interact with scientists, researchers, potential research subjects, practitioners, and other stakeholders in policy development. An assessment based on intuitive ethics combined with stakeholder perspectives for CRISPR and gene drive technologies needs to be promoted and developed. With the development and use of CRISPR and gene drive technology, it is important to identify what is being valued in their use and what will be valued in how they will be used in the future, outside of the scope of the current technology.

Thus, single parameter evaluations (from the perspective of a discipline or a single stakeholder) distort the complex problems and set aside valuable considerations crucial for social justice and the global public good. It is important to see when ethical considerations need to play a role in developing public policy. It is essential to recognize the need to engage in ethical analysis even when it may matter little to public policy. Although some may argue that this technology is not developed enough for a policy discussion, we believe that an AEA of the current technology concerning possible outcomes and future persons is paramount given its rapid development and implementation.

This work aims to focus ethical reflection using our categories by those who engage in the development and application of these technologies, focusing on the long-term consequences for all the stakeholders who will potentially be affected by the technology. As we articulated in Chap. 5, there is a high social cost when the lay public is not engaged in the CRISPR debate at early stages (Jasanoff et al., 2015). When we don't include all possible stakeholders, including the perspective of future persons, we assume that our successors will agree with our intentions, our actions, and what we saw as short- and long-term consequences. In the case of CRISPR and gene drive in humans, our AEA attempts to curtail our scientific hubris by exploring the viewpoints of all the stakeholders involved. By doing so, we hope to increase justice, freedom, and, ultimately, happiness for current and future individuals in the spirit of Audi.

References

Akcakaya, P., Bobbin, M. L., Guo, J. A., et al. (2018). In vivo CRISPR editing with no detectable genome-wide off-target mutations. *Nature, 561*(7723), 416–419. https://doi.org/10.1038/s41586-018-0500-9

Allhoff, F., Henschke, A., & Strawser, B. J. (2016). *Binary bullets: The ethics of cyberwarfare*. Oxford University Press.

Andreasen, N. C. (2008). The relationship between creativity and mood disorders. *Dialogues in Clinical Neuroscience, 10*(2), 251–255.

Audi, R. (2007). *Moral value and human diversity*. Oxford University Press.

Babar, M. I., Ghazali, M., Jawawi, D. N., & Bin Zaheer, K. (2015). StakeMeter: Value-based stakeholder identification and quantification framework for value-based software systems. *PLoS One, 10*(3), e0121344.

Ben Ouagrham-Gormley, S., & Vogel, K. M. (2016, October 14). Gene drives: The good, the bad, and the hype. *Bulletin of the Atomic Scientists*. http://thebulletin.org/gene-drives-good-bad-and-hype10027. Accessed 2 Aug 2018.

Bogardus, K., & Saplakoglu, Y. (2017, February 19). New gene drive technology could wipe out malaria, but is it safe? *Science Blog*. http://www.sciencemag.org/news/2017/02/new-gene-drive-technology-could-wipe-out-malaria-it-safe. Accessed 2 Aug 2018.

Cuilla, J., Martin, C., & Solomon, R. C. (2007). *Honest business, a business ethics reader*. Oxford University Press.

De Wert, G., Heindryckx, B., Pennings, G., Clarke, A., Eichenlaub-Ritter, U., Van El, C. G., Cornel, M. C. (2018). Responsible innovation in human germline gene editing: Background document to the recommendations of ESHG and eshre. *European Journal of Human Genetics, 26*(4), 450–470. https://doi.org/10.1038/s41431-017-0077-z

Esvelt, K. M., Smidler, A. L., Catteruccia, F., & Church, G. M. (2014). Concerning RNA-guided gene drives for the alteration of wild populations. *eLife, 3*, e03401.

Fu, Y., Foden, J. A., Khayter, C., et al. (2013). High-frequency off-target mutagenesis induced by CRISPR-Cas nucleases in human cells. *Nature Biotechnology, 31*(9), 822–826.

Gardner, H. (2017, May 19). What can CRISPR do to help in the fight against Zika virus? *The Bench, Official Synthego Blog*. https://www.synthego.com/blog/what-can-crispr-do-to-help-in-the-fight-against-zika-virus/. Accessed 2 Aug 2018.

Gulati, S. (2008). Technology-enhanced learning in developing nations: A review. *The International Review of Research in Open and Distributed Learning, 9*(1), 1–8.

Jasanoff, S., Hurlbut, J. B., & Saha, K. (2015). CRISPR democracy: Gene editing and the need for inclusive deliberation. *Issues in Science and Technology, 32*.

Ma, H., Marti-Gutierrez, N., Park, S. W., et al. (2017). Correction of a pathogenic gene mutation in human embryos. *Nature, 548*(7668), 413–419. https://doi.org/10.1038/nature23305

National Academies of Sciences, Engineering, and Medicine. 2016. Gene Drives on the Horizon: Advancing Science, Navigating Uncertainty, and Aligning Research with Public Values. Washington, DC: The National Academies Press. https://doi.org/10.17226/23405.

Nestor, M. W., & Wilson, R. L. (2018). Can the use of CRISPR in humans result in decreased social justice for future stakeholders? *Ethics in Biology, Engineering and Medicine: An International Journal, 9*, 5–16. https://doi.org/10.1615/EthicsBiologyEngMed.2019030551

Nestor, M. W., & Wilson, R. L. (2020). Beyond Mendelian genetics: Anticipatory biomedical ethics and policy implications for the use of CRISPR together with gene drive in humans. *Journal of Bioethical Inquiry, 17*(1), 133–144. https://doi.org/10.1007/s11673-019-09957-7

Polcz, S., & Lewis, A. (2016). CRISPR-Cas9 and the non-germline non-controversy. *Journal of Law and the Biosciences, 3*(2), 413–425. https://doi.org/10.1093/jlb/lsw016

Schwank, G., Koo, B. K., Sasselli, V., et al. (2013). Functional repair of CFTR by CRISPR/Cas9 in intestinal stem cell organoids of cystic fibrosis patients. *Cell Stem Cell, 13*(6), 653–658.

Sun, N., & Zhao, H. (2014). Seamless correction of the sickle cell disease mutation of the HBB gene in human induced pluripotent stem cells using TALENs. *Biotechnology and Bioengineering, 111*(5), 1048–1053.

Chapter 7
Future Use of CRISPR II: CRISPR Kill Switches

7.1 Introduction

Although in Chap. 1 we tried to show the promise and potential of CRISPR to enhance our ability to reduce suffering and treat disease, a large portion of this manuscript has been devoted to potential negative consequences of the use of CRISPR in specific contexts. It is essential then to rebalance the discussion by reviewing current scientific attempts to guardrail CRISPR and gene drive (we will discuss policy attempts to guardrail in the next chapter).

The initial discovery of what is now called the CRISPR locus in 1993 was followed by decades of research on the CRISPR-Cas9 system to better characterize and understand its mechanism and functionality for genome editing (Ishino et al., 2018). Unsurprisingly, this novel finding has resulted in a drastic influx of preclinical studies utilizing the CRISPR system, with the intent of ameliorating pathologies that plague the modern world. However, as with numerous other new biotechnologies that have the potential for use in a clinical setting, CRISPR's side effects and safety concerns have and are still being continuously identified. For example, a major point of concern for the use of the CRISPR-Cas9 biotechnology is its long-term safety, specifically regarding the potential tumorigenesis that can result if off-target gene editing occurs on oncogenes or tumor-suppressor genes. Other significant side effects recently identified and associated with CRISPR-Cas9 are the unexpected chromosomal truncations and/or deletions of large DNA fragments after cleavage (Adikusuma et al., 2018; Fu et al., 2013).

The safety concerns outlined above have sparked innovation amongst the scientific community with the development of anti-CRISPR (Acr) proteins, or CRISPR 'kill switches' which enhance the control and precision of the CRISPR-Cas9 biotechnology, presumably mitigating the system's side effects. The proteins were first discovered in the early 2010s by infecting different bacteriophages into the common laboratory bacterial strain Pseudomonas aeruginosa PA14. Researchers found that when bacteriophages, which naturally express anti-CRISPR (AcrIF1-5) proteins,

M. W. Nestor, R. L. Wilson, *Anticipatory Ethics and The Use of CRISPR in Humans*, https://doi.org/10.1007/978-3-030-98368-0_7

underwent mutation of these anti-CRISPR genes, they could not infect bacteria with a functional CRISPR/Cas9 system, such as PA14 (Bondy-Denomy et al., 2013). The subsequent addition of the Acr gene back into the genome of bacteriophages enabled them to evade the CRISPR/Cas9 system, allowing them to replicate and infect the host bacteria. Interestingly, further investigation has discovered that the innate bacterial CRISPR/Cas9 system has a mechanism by which it can 'retarget' mutated bacteriophages to maintain resistance (Cady et al., 2012), and reports have shown that it possesses adaptive capabilities based on nucleotides within the protospacer adjacent motif (PAM) nucleotide sequence. Together, these studies provided novel insight concerning the adaptability of the CRISPR/Cas9 system and the existence of inherent anti-CRISPR proteins expressed in different viruses.

Since the discovery of the viral anti-CRISPR capabilities and nine anti-CRISPR protein families closely related to P. aeruginosa bacteriophages, over 80 distinct families of anti-CRISPR proteins have been identified and published. An anti-CRISPR database and classification scheme have been made to register such discoveries (Bondy-Denomy et al., 2018). These findings have provided the basis for anti-CRISPR protein research, which aims to mitigate the side effects and off-target deletions previously observed in pre-clinical investigations, ultimately by increasing control and precision of the CRISPR/Cas9 editing process using these proteins as spatiotemporal 'kill switches' of Cas9 activity. Thus, the purpose of the following text is to provide a scientific review of some of the major phage-derived anti-CRISPR 'kill switches', their efficacy, utility, and overall impact in pre-clinical in vitro and/or in vivo settings.

7.2 Anti-CRISPR Proteins That Block DNA Binding

Currently, the most common mechanism by which anti-CRISPR proteins can evade the CRISPR/Cas9 system is through their ability to block DNA binding, and eight of the eleven anti-CRISPR proteins with known mechanisms have displayed this property (Hwang & Maxwell, 2019). Many of the anti-CRISPR proteins achieve this blocking through different methods as well; for instance, AcrIF1, AcrIF2, and AcrIF10 hinder the type I-F CRISPR-Cas system by preventing DNA binding to the Cascade effector complex using different mechanisms. Specifically, AcrIF1 binds the Cas7f backbone of the Cascade effector complex, inducing a conformational change that thwarts the target DNA from binding, accessing, and ultimately hybridizing with the CRISPR RNA (crRNA), a step that is crucial for the CRISPR/Cas system to effectively cleave DNA targets (Chowdhury et al., 2017). These mechanisms were demonstrated through size-exclusion chromatography and cryo-electron microscopy structural analyses (Guo et al., 2017; Peng et al., 2017; Bondy-Denomy et al., 2015). AcrIF2 exhibits its blocking properties using a different mechanism—the anti-CRISPR protein can block DNA target binding by wedging itself in between the Cas7f subunit on the tail end of the DNA backbone and Cas8f, allowing it to compete with the Cas enzymes for binding to critical DNA sites (Guo et al., 2017;

Peng et al., 2017). Finally, AcrIF10 has shown that it can force the effector Cascade complex to adopt the DNA-bound conformation by binding to a specific binding site formed by Cas8f and Cas5f, inhibiting effective CRISPR/Cas cleavage as well (Guo et al., 2017) (Fig. 7.1).

The type II CRISPR/Cas system is characterized by the utilization of the Cas9 effector Cascade, a double-stranded DNA target (dsDNA), pre-crRNA processing using RNase III, the use of trans-activating CRISPR RNA (tracrRNA), and a 3'G-rich PAM sequence; and a few anti-CRISPR proteins have displayed DNA binding properties against this system specifically to block and inhibit it (Tang & Fu, 2018). For example, cryo-electron microscopy observations have shown that AcrIIC3 can dimerize Type II-C Cas9, preventing the ability to bind its DNA target (Harrington et al., 2017). Further studies using electrophoretic mobility shift assays have shown that AcrIIC4 and AcrIIC5 can inhibit DNA binding of another Cas9 ortholog from N. meningitidis, NmeCas9, even after the single guide RNA (sgRNA) is loaded into the NmeCas9 complex (Lee et al., 2018). Microscopy and structural analyses identified that AcrIIA2 could act as a DNA mimic, inhabiting the DNA binding site on Cas9 completely to inhibit Cas9 DNA binding (Jiang et al., 2019). Lastly, crystal structure analysis of AcrIIA4-bound SpyCas9, the most commonly used gene-editing CRISPR/Cas9 system from Streptococcus pyogenes strain SF370 (Jinek et al., 2012) has provided a new mechanism for ACRIIA4 as an anti-CRISPR protein as well. The report found that ACRIIA4 structurally mimics the innate PAM sequence, allowing it to bind the PAM-interacting site of the PAM-interacting domain, thereby blocking recognition of dsDNA substrates by SpyCas9 (Dong et al., 2017). Interestingly, AcrIIA4 also obstructs SpyCas9 endonuclease activity by shielding its RuvC active site from cleavage.

Altogether, these technical studies demonstrate the diverse mechanisms by which the anti-CRISPR proteins can evade the CRISPR/Cas system. Moreover, while most of these mechanisms involve blocking the DNA in various ways, other mechanisms for evasion have been described. One example of this is the prevention of DNA cleavage, a property displayed by three anti-CRISPR proteins: AcrIE1, AcrIF3, and AcrIIC1 (Hwang & Maxwell, 2019). AcrIE1 can bind Cas3 in type I CRISPR/Cas systems, inhibiting Cas3's ability to cleavage target DNA (Pawluk et al., 2017); while AcrIF3 displays a direct Cas3 binding characteristic to impede recruitment of Cas3 to the Cascade complex, thereby preventing DNA cleavage as well (Bondy-Denomy et al., 2015). In addition, X-ray crystallography and cryo-electron microscopy structures established that AcrIF3 binds to Cas3 as a dimer, rendering blockage of Cas3 to the Cascade complex as well (Wang et al., 2016). Finally, only one anti-CRISPR protein, AcrIIC1, has demonstrated the ability to prevent DNA cleavage mediated by the type II CRISPR/Cas9 system. Biochemical and structural analyses revealed that AcrIIC1 could bind directly to the Cas9 HNH endonuclease domain, inhibiting DNA cleavage while allowing the target DNA to bind Cas9, resulting in a 'dead' Cas9 (dCas9) (Harrington et al., 2017).

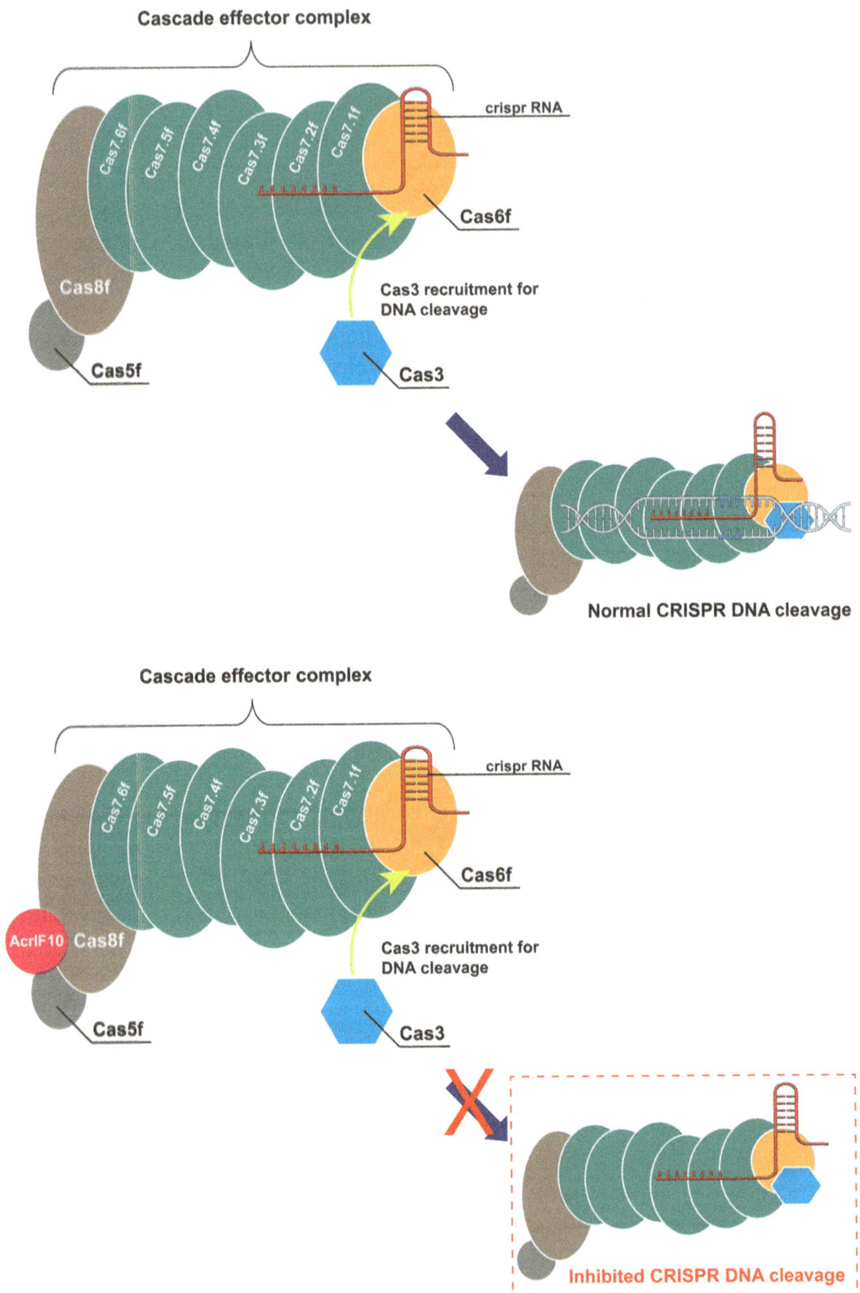

Fig. 7.1 There are several anti-CRISPR approaches, including using proteins that inhibit Cas protein DNA binding, thereby inhibiting the activity of Cas and CRISPR-mediated gene editing. One example is the use of Acr-binding to the Cascade effector complex. AcrIF10, when associated with the Cascade effector complex by binding to a site associated with Cas5f and Cas8f, induces a change in the conformation of the complex. This alteration of the conformation inhibits DNA cleavage by Cas and thus inhibits CRISPR-editing

7.3 Artificial Anti-CRISPR Proteins

Despite the aforementioned utility of anti-CRISPR proteins, the current, most common limitation is that current methods to easily restrict the time and locations of Acr activity on Cas9 regulation are lacking. Because of this, studies have focused on engineering artificial anti-CRISPR proteins to avoid these issues. For example, a 2018 study utilized the light, oxygen, or voltage (LOV2) domain from A. sativa phototropin-1, which, after insertion into protein, undergoes protein unfolding in response to photoexcitation (blue light), resulting in impairments in protein function as long as the LOV2 insertion loop is conformationally coupled to an enzyme's active site (Bubeck et al., 2018; Dagliyan et al., 2016). The researchers inserted this LOV2 domain into an 87-amino-acid anti-CRISPR protein called AcrIIA4, derived from the prophage Listeria monocytogenes. Insertion resulted in an Acr-LOV hybrid that exhibited clear, potent associations with Cas9 activity: effective inhibition of Cas9 activity in the dark and recovery of Cas9 activity after photoactivation, as identified by a luciferase-reporter cleavage assay (Bubeck et al., 2018). The most effective of these hybrids was one termed CASANOVA, which carried a three-amino-acid deletion preceding the LOV domain insertion site; and, when CASANOVA was transfected into human embryonic kidney cells (HEK293T) or human bone osteosarcoma epithelial cells (U2OS), Cas9 activity reacted similarly in response to changes in light exposure. These experiments provide evidence for the anti-CRISPR AcrIIA4 protein as one of the major players involved in Listeria monocytogenes integration into bacteria and its ability to bind Cas9 and inhibit its activity. The report also adds another technique to better understand anti-CRISPR proteins through the use of hybrid proteins that display CRISPR/Cas9 regulatory activity in response to photoexcitation.

7.4 Anti-CRISPR Associated Genes

The acr genes translate into the Acr proteins described above and have previously been characterized by highly variable and non-homologous genetic material (Li & Bondy-Denomy, 2021). Interestingly, recent discoveries in the past 2 years have found functional importance for a set of conserved genes that encode predicted helix-turn-helix (HTH) DNA-binding domains next to these acr genes and have since been termed 'anti-CRISPR-associated' genes, or aca. One of the first experiments involved in this discovery involved using the P. aeruginosa bacteriophage JBD30, which comprises an acrIF1-aca1 locus and rapidly expresses acrIF1 following infection into bacteria (Stanley et al., 2019). In response to accumulation of Aca1, expression of the AcrIF1 protein is diminished, acting as a form of negative feedback; and deletion of aca1 from the bacteriophage genome resulted in toxic overexpression of Acr1, decreased transcription of downstream genes essential to phage proliferation and, eventually, phage lethality (Stanley et al., 2019). Similarly,

Pectobacterium carotovorum exhibits this negative feedback role with aca2, and Neisseria with aca3 (Birkholz et al., 2019; Stanley et al., 2019). These data indicate a novel mechanism of the anti-CRISPR proteins by which neighboring genes, or anti-CRISPR associated genes, control their expression for optimal proliferation.

7.5 Defense Advanced Research Projects Agency (DARPA) : Safe Genes Project

With CRISPR biotechnology emerging as an important tool for gene editing over the past few years and significant clinical implications in the future, DARPA created and invested in a 4 year, $65 million Safe Genes program in 2017, a program encompassing seven teams from various prestigious locations, including The Broad Institute of MIT and Harvard, Harvard Medical School, Massachusetts General Hospital (MGH), Massachusetts Institute of Technology (MIT), North Carolina State University, University of California-Berkeley (UCB), and University of California Riverside. The program's technical objectives entail the development of CRISPR/Cas genetic constructs, the control and utility of CRISPR in living systems, and new methods for effective drug-based treatment strategies that garnish the benefits of CRISPR technology while still preserving genome integrity. Safe Genes program manager, Renee Wegrzyn, commented on the first phase of the Safe Genes program, stating that it initially focused on the "technological foundations and early proofs of concept to determine which research pathways show the most promise." (DARPA, 2019)

Following what Renee Wegrzyn describes as phase I and the inception of DARPA's program, the teams have contributed to significant advancements towards the CRISPR field. The MGH demonstrated new CRISPR design technology that improves gene editing capabilities using a multifaceted approach to increase control and accuracy on CRISPR/Cas targets (Akcakaya et al., 2018). In the report, the MGH team discusses this innovative technology called VIVO, or Verification of In Vivo Off targets, which displayed the efficacy of and improvements of their CRISPR/Cas experiments in laboratory experiments on different mouse models. Another significant contribution by the MGH team has been the software resource 'CRISPResso2,' an open-source software that can be used to rapidly detect and analyze genome editing sequences (Clement et al., 2019). The development of CRISPResso2 has made genome editing analysis significantly more straightforward and quicker, as the high-throughput sequencing technology can process data much more accurately and quickly than before (DARPA, 2019). Additionally, the DARPA-funded MGH team has also engineered a structure-guided base editor, enhancing the control and specificity of CRISPR/Cas transcript editing in human cells, and DARPA has stated that this research "is an important first step toward translating CRISPR-Cas genome editing tools from laboratory use into clinical use." (DARPA, 2019)

Other teams involved in the DARPA project are actively developing novel genome editing tools as well, with a significant emphasis on improving CRISPR/Cas specificity to protect the genomic integrity of organisms against nonspecific changes. This research, conducted by UCB and MGH, has resulted in the first identifications of type I and type V anti-CRISPR proteins, which display specific blocking capabilities against Cas12 to evade gene editing in human cell culture (Marino et al., 2018; Watters et al., 2018). Furthermore, complementing the above anti-CRISPR discoveries was the work led by the Broad Institute's Safe Genes team, which has established novel high-throughput methods to detect Cas proteins and inhibitors to identify new genome editing regulators (Maji et al., 2019; Manna et al., 2019).

Overall, the Safe Genes program by DARPA has led to significant CRISPR/Cas innovation by some of the leading investigators throughout the United States. New mechanisms of CRISPR/Cas function and anti-CRISPR proteins have been identified that can potentially be used for clinical application in the future, along with publicly available resources that enable researchers worldwide to quickly and accurately analyze sequencing data for CRISPR/Cas target and off-target sites. DARPA and the Safe Genes program have made it a point to emphasize the goal of their research, which is to find ways to limit the use of genome editing tools to preserve the organism's genomic integrity and protect it from unwanted alterations. This aspect of CRISPR/Cas9 research needs to be considered a fundamental goal of other programs in the community, as mitigating the risks and security concerns, either accidental or intentional, is vital for potential use in the clinical setting.

References

Adikusuma, F., Piltz, S., Corbett, M. A., Turvey, M., Mccoll, S. R., Helbig, K. J., Beard, M. R., Hughes, J., Pomerantz, R. T., & Thomas, P. Q. (2018). Large deletions induced by Cas9 cleavage. *Nature, 560*, E8–E9.

Akcakaya, P., Bobbin, M. L., Guo, J. A., Malagon-Lopez, J., Clement, K., Garcia, S. P., Fellows, M. D., Porritt, M. J., Firth, M. A., Carreras, A., Baccega, T., Seeliger, F., Bjursell, M., Tsai, S. Q., Nguyen, N. T., Nitsch, R., Mayr, L. M., Pinello, L., Bohlooly-Y, M., … Joung, J. K. (2018). In vivo CRISPR editing with no detectable genome-wide off-target mutations. *Nature, 561*, 416–419.

Birkholz, N., Fagerlund, R. D., Smith, L. M., Jackson, S. A., & Fineran, P. C. (2019). The autoregulator Aca2 mediates anti-CRISPR repression. *Nucleic Acids Research, 47*, 9658–9665.

Bondy-Denomy, J., Davidson, A. R., Doudna, J. A., Fineran, P. C., Maxwell, K. L., Moineau, S., Peng, X., Sontheimer, E. J., & Wiedenheft, B. (2018). A unified resource for tracking anti-CRISPR names. *The CRISPR Journal, 1*, 304–305.

Bondy-Denomy, J., Garcia, B., Strum, S., Du, M., Rollins, M. F., Hidalgo-Reyes, Y., Wiedenheft, B., Maxwell, K. L., & Davidson, A. R. (2015). Multiple mechanisms for CRISPR–Cas inhibition by anti-CRISPR proteins. *Nature, 526*, 136–139.

Bondy-Denomy, J., Pawluk, A., Maxwell, K. L., & Davidson, A. R. (2013). Bacteriophage genes that inactivate the CRISPR/Cas bacterial immune system. *Nature, 493*, 429–432.

Bubeck, F., Hoffmann, M. D., Harteveld, Z., Aschenbrenner, S., Bietz, A., Waldhauer, M. C., Börner, K., Fakhiri, J., Schmelas, C., Dietz, L., Grimm, D., Correia, B. E., Eils, R., & Niopek,

D. (2018). Engineered anti-CRISPR proteins for optogenetic control of CRISPR-Cas9. *Nature Methods, 15*, 924–927.

Cady, K. C., Bondy-Denomy, J., Heussler, G. E., Davidson, A. R., & O'toole, G. A. (2012). The CRISPR/Cas adaptive immune system of Pseudomonas aeruginosa mediates resistance to naturally occurring and engineered phages. *Journal of Bacteriology, 194*, 5728–5738.

Chowdhury, S., Carter, J., Rollins, M. F., Golden, S. M., Jackson, R. N., Hoffmann, C., Nosaka, L., Bondy-Denomy, J., Maxwell, K. L., Davidson, A. R., Fischer, E. R., Lander, G. C., & Wiedenheft, B. (2017). Structure reveals mechanisms of viral suppressors that intercept a CRISPR RNA-guided surveillance complex. *Cell, 169*, 47–57.e11.

Clement, K., Rees, H., Canver, M. C., Gehrke, J. M., Farouni, R., Hsu, J. Y., Cole, M. A., Liu, D. R., Joung, J. K., Bauer, D. E., & Pinello, L. (2019). CRISPResso2 provides accurate and rapid genome editing sequence analysis. *Nature Biotechnology, 37*, 224–226.

Dagliyan, O., Tarnawski, M., Chu, P. H., Shirvanyants, D., Schlichting, I., Dokholyan, N. V., & Hahn, K. M. (2016). Engineering extrinsic disorder to control protein activity in living cells. *Science, 354*, 1441–1444.

Darpa. (2019). Safe genes tool kit takes shape [Online]. *DARPA*. Available: https://www.darpa.mil/news-events/2019-10-15. Accessed 28 July 2021.

Dong, D., Guo, M., Wang, S., Zhu, Y., Wang, S., Xiong, Z., Yang, J., Xu, Z., & Huang, Z. (2017). Structural basis of CRISPR–SpyCas9 inhibition by an anti-CRISPR protein. *Nature, 546*, 436–439.

Fu, Y., Foden, J. A., Khayter, C., Maeder, M. L., Reyon, D., Joung, J. K., & Sander, J. D. (2013). High-frequency off-target mutagenesis induced by CRISPR-Cas nucleases in human cells. *Nature Biotechnology, 31*, 822–826.

Guo, T. W., Bartesaghi, A., Yang, H., Falconieri, V., Rao, P., Merk, A., Eng, E. T., Raczkowski, A. M., Fox, T., Earl, L. A., Patel, D. J., & Subramaniam, S. (2017). Cryo-EM structures reveal mechanism and inhibition of DNA targeting by a CRISPR-Cas surveillance complex. *Cell, 171*, 414–426.e12.

Harrington, L. B., Doxzen, K. W., Ma, E., Liu, J. J., Knott, G. J., Edraki, A., Garcia, B., Amrani, N., Chen, J. S., Cofsky, J. C., Kranzusch, P. J., Sontheimer, E. J., Davidson, A. R., Maxwell, K. L., & Doudna, J. A. (2017). A broad-spectrum inhibitor of CRISPR-Cas9. *Cell, 170*, 1224–1233.e15.

Hwang, S., & Maxwell, K. L. (2019). Meet the anti-CRISPRs: Widespread protein inhibitors of CRISPR-Cas systems. *The CRISPR Journal, 2*, 23–30.

Ishino, Y., Krupovic, M., & Forterre, P. (2018). History of CRISPR-Cas from encounter with a mysterious repeated sequence to genome editing technology. *Journal of Bacteriology, 200*, e00580–e00517.

Jiang, F., Liu, J. J., Osuna, B. A., Xu, M., Berry, J. D., Rauch, B. J., Nogales, E., Bondy-Denomy, J., & Doudna, J. A. (2019). Temperature-responsive competitive inhibition of CRISPR-Cas9. *Molecular Cell, 73*, 601–610.e5.

Jinek, M., Chylinski, K., Fonfara, I., Hauer, M., Doudna, J. A., & Charpentier, E. (2012). A programmable dual-RNA–guided DNA endonuclease in adaptive bacterial immunity. *Science, 337*, 816–821.

Lee, J., Mir, A., Edraki, A., Garcia, B., Amrani, N., Lou, H. E., Gainetdinov, I., Pawluk, A., Ibraheim, R., Gao, X. D., Liu, P., Davidson, A. R., Maxwell, K. L., Sontheimer, E. J., Charpentier, E., Barrangou, R., & Moineau, S. (2018). Potent Cas9 inhibition in bacterial and human cells by AcrIIC4 and AcrIIC5 anti-CRISPR. *Proteins, mBio, 9*, e02321–e02318.

Li, Y., & Bondy-Denomy, J. (2021). Anti-CRISPRs go viral: The infection biology of CRISPR-Cas inhibitors. *Cell Host & Microbe, 29*, 704–714.

Maji, B., Gangopadhyay, S. A., Lee, M., Shi, M., Wu, P., Heler, R., Mok, B., Lim, D., Siriwardena, S. U., Paul, B., Dančík, V., Vetere, A., Mesleh, M. F., Marraffini, L. A., Liu, D. R., Clemons, P. A., Wagner, B. K., & Choudhary, A. (2019). A high-throughput platform to identify small-molecule inhibitors of CRISPR-Cas9. *Cell, 177*, 1067–1079.e19.

Manna, D., Maji, B., Gangopadhyay, S. A., Cox, K. J., Zhou, Q., Law, B. K., Mazitschek, R., & Choudhary, A. (2019). A singular system with precise dosing and spatiotemporal control of CRISPR-Cas9. *Angewandte Chemie International Edition, 58*, 6285–6289.

Marino, N. D., Zhang, J. Y., Borges, A. L., Sousa, A. A., Leon, L. M., Rauch, B. J., Walton, R. T., Berry, J. D., Joung, J. K., Kleinstiver, B. P., & Bondy-Denomy, J. (2018). Discovery of widespread type I and type V CRISPR-Cas inhibitors. *Science, 362*, 240–242.

Pawluk, A., Shah, M., Mejdani, M., Calmettes, C., Moraes, T. F., Davidson, A. R., Maxwell, K. L., & Charpentier, E. (2017). Disabling a type I-E CRISPR-Cas nuclease with a bacteriophage-encoded anti-CRISPR protein. *MBio, 8*, e01751–e01717.

Peng, R., Xu, Y., Zhu, T., Li, N., Qi, J., Chai, Y., Wu, M., Zhang, X., Shi, Y., Wang, P., Wang, J., Gao, N., & Gao, G. F. (2017). Alternate binding modes of anti-CRISPR viral suppressors AcrF1/2 to Csy surveillance complex revealed by cryo-EM structures. *Cell Research, 27*, 853–864.

Stanley, S. Y., Borges, A. L., Chen, K.-H., Swaney, D. L., Krogan, N. J., Bondy-Denomy, J., & Davidson, A. R. (2019). Anti-CRISPR-associated proteins are crucial repressors of anti-CRISPR transcription. *Cell, 178*, 1452–1464.e13.

Tang, Y., & Fu, Y. (2018). Class 2 CRISPR/Cas: An expanding biotechnology toolbox for and beyond genome editing. *Cell & Bioscience, 8*, 59.

Wang, J., Ma, J., Cheng, Z., Meng, X., You, L., Wang, M., Zhang, X., & Wang, Y. (2016). A CRISPR evolutionary arms race: Structural insights into viral anti-CRISPR/Cas responses. *Cell Research, 26*, 1165–1168.

Watters, K. E., Fellmann, C., Bai, H. B., Ren, S. M., & Doudna, J. A. (2018). Systematic discovery of natural CRISPR-Cas12a inhibitors. *Science, 362*, 236–239.

Chapter 8
Domestic and International Regulation of CRISPR

8.1 Introduction

While CRISPR technology has shown that it can have a profound impact on an individual's life concerning treating a variety of serious diseases, its ability as a genomic editing tool for prevention and treatment, as well as many other innovative, biological applications, pose the question of the ethics behind CRISPR concerning international human rights, and what restrictions should come along with it. As others have eloquently discussed previously, human genomic modifications are at the very least heavily restricted, and some even prohibited (van Beers, 2020; Yotova, 2020). As genomic editing is a new and growing topic concerning its clinical applications in humans, discussions on these topics are ongoing, and a lack of a clear definition of the ethics of CRISPR and how tightly it should be regulated exists. Thus, this chapter will provide a framework for domestic and international regulation of genomic editing and CRISPR, including background and recent CRISPR developments, the potential legal issues that may arise from it, and how CRISPR is being regulated today.

8.2 Background and Recent Developments

News in November of 2018 triggered possibly one of the most controversial and ethical events in the CRISPR space. The Chinese scientist Dr. He Jiankui went on YouTube to declare the birth of two healthy genetically modified babies named Lulu and Nana (Lab, 2018). Dr. Jiankui had used the CRISPR genome editing technique to modify embryonic DNA and implanted these embryos for pregnancy, an experiment conducted in a private Chinese hospital that entirely violated the country's policy on guidelines that prohibited genomic editing of the kind. Dr. Jiankui directed this experiment in eight couples with HIV-positive fathers, intending to make the

M. W. Nestor, R. L. Wilson, *Anticipatory Ethics and The Use of CRISPR in Humans*, https://doi.org/10.1007/978-3-030-98368-0_8

genetically engineered embryonic offspring resistant to the virus by targeting a gene known as C-C chemokine receptor type 5, or CCR5 ("World's First Gene-Edited Babies Created in China," 2018). The announcement was received poorly, and soon after that, the World Health Organization stated the governance and oversight of genomic editing in humans, which included universal distaste of the experiment by scientists and international organizations who reiterated the belief that clinical research utilizing genomic editing was unacceptable at the time (WHO, 2019).

Moreover, the Second International Summit on Human Genome Editing initially announced Dr. Jiankui's experiment. They mentioned that "the procedure was irresponsible and failed to conform with international norms" (Statement by the Organizing Committee of the Second International Summit on Human Genome Editing, 2018). This declaration was consistent with numerous other international concerns, including those expressed by the UK Nuffield Council on Bioethics, who in their article reported that over 100 Chinese biomedical researchers strongly condemned the CRISPR investigation and a consensus on genome editing needs to be met (Cyranoski & Ledford, 2018). Furthermore, the French National Academy of Medicine also issued a national report on the subject reproving the investigation and noting that the experiment raised many unresolved scientific, medical, and ethical issues (A declaration of the Académie nationale de médecine and of the Académie des sciences concerning the announcement made by Dr Hé Jiangkui, 2018). Finally, multiple Chinese organizations, including the Chinese Academy of Engineering and the Chinese Academy of Medical Sciences, expressed their concerns in correspondence, highlighting worldwide concern sparked by the experiment done by Dr. Jiankui (Chinese Academy of Medical Sciences: "Gene-edited Babies: Chinese Academy of Medical Sciences' Response and Action," 2018).

Unsurprisingly, news of the CRISPR experiment ignited concerns across the globe given the lack of formidable research and the fact that the technology is still in experimental stages, which prompted questions about the status of genome editing and its regulation under international law. Furthermore, Dr. Jiankui's experiment did not achieve success in producing offspring with resistance to HIV, produced off-target genetic alterations, and novel mutations of the CCR5 gene (which confers natural HIV resistance) were identified, highlighting the inconsistencies of the newly used, highly technical genomic editing methodology (Greely, 2019). With this in mind, an inventor of the CRISPR-Cas9 system, Jennifer Doudna, commented on such disregard for scientific ethicality with her statement:

> He's fateful decision to ignore the basic medical mantra of "do no harm" and risk the unintended consequences will likely be remembered as one of the most shocking misapplications of any scientific tool in our history…

Doudna addressed the necessity for the scientific community to quickly establish stronger safeguards for genomic editing in research (Doudna, 2019). On December 30, 2019, He was sentenced to 3 years of imprisonment by the local Chinese court for committing "Illegal Medical Practice," along with two other researchers on his team that were also sentenced.

Importantly, genomic editing in humans and the implications of these technologies on reproduction are not exactly new; other reproductive machinery, including noninvasive prenatal testing (NIPT) and reimplantation genetic diagnosis (PGD), had previously made it possible to give parents a sort of genetic selection of their type of child in the 1990s and beyond. In 2016, human nuclear genome transfer (HNGT), a non-CRISPR genome modifying technique known as mitochondrial replacement therapy was carried out by a Chinese-American fertility doctor named John Zhang, which resulted in the first "three parent" birth of a genetically modified Jordanian baby in Mexico (Reardon, 2017). The development of CRISPR and other genome-altering methodologies has made it possible to genetically modify offspring, rendering it conceivable to override genetic "lottery" outcomes and generate an entirely new set of ethical dilemmas. With these concerns in mind, the question then is what were the major international laws governing CRISPR and human genomic editing as a whole before the birth of these children, and how were these laws changed in response? These laws and regulations will be outlined in the following chapter, paying particular respect to the regulatory framework concerning genetic alterations of human embryonic and gametic DNA, a modification commonly referred to as human genomic germline editing as it involves genetic modification germ cells.

8.3 National Regulation of Human Genomic Modifications

Currently, the legal framework and regulatory stance concerning the use of CRISPR and any human genome editing biotechnologies in the human germline is that this use should have heavy restrictions or be entirely banned. In a 2014 review on the regulation of human genomic germline editing, which aimed to outline the national ethical issues involved with clinical genomic germline editing and whether it should be permitted for use in preventative medicine, Araki and Ishii articulated that a large majority of countries with such regulations completely banned such technology (29 of 39 countries), while nine of these countries had overly ambiguous policies (Araki & Ishii, 2014). Notably, 29 countries have also ratified the Council of Europe's *Convention for the Protection of Human Rights and Dignity of the Human Being with regard to the Application of Biology and Medicine: Convention on Human Rights and Biomedicine* (the Oviedo Convention), which took place in Oviedo, Spain (Council, 1997; Hendriks, 1997). Discussions at the Oviedo Convention and approval of the guidelines by these 29 countries resulted in the prohibition of human genome germline editing, with the caveat of Article 13, which stated that

> an intervention seeking to modify the human genome may only be undertaken for preventative, diagnostic or therapeutic purposes and only if its aim is not to introduce any modification in the genome of any descendants.

These ratifications legally bound each of the 29 countries to the policies laid out in the Oviedo Conventions regulations and detailed that CRISPR and other human genomic germline-editing technologies were not entirely banned as specific research and diagnostic scenarios are allowed.

The previous review on human genomic germline editing on those 39 countries left only one country, the USA, as one of the only nations in 2014 without a complete ban on the technology but severely restricted guidelines (Liu, 2020); however, as the aforementioned text alluded to, Article 13 of the Oviedo Convention legislation did suggest a complete ban on genomic germline editing technologies was not enforced in many European countries as well. A further examination of the regulations was done in 2016, when human genomic germline editing guidelines continued to become increasingly complex. The degree of these national bans and restrictions began to vary significantly, as some countries, including much of the European Union, Australia, Canada, and Brazil, accompanied the prohibition of human genome editing with criminal sanctions. In contrast, other countries allowed CRISPR and other human genome editing technologies for research (Isasi et al., 2016). However, in some of the perhaps more lenient national regulations in countries like the USA, the UK, and China, laws and regulations were, and are, still in place, which imposes stringent limits with research use of human genome editing technologies (van Beers, 2020).

8.4 National Regulation in the USA

More specifically, in the USA, human genome editing and CRISPR particularly are not banned. Still, strict rules exist under regulation by the Food and Drug Administration (FDA), which regulates clinical studies in the USA, along with a set of guidelines provided by the National Institutes of Health (NIH) (Liu, 2020). These rules and regulations are particularly important in the USA, as the country produces the most research involving basic genomic editing ("Nuffield Council,"). Of the guidelines provided by the NIH, they have made it clear that any clinical trials proposed which utilize germline genomic editing alterations will be rejected by the NIH's Recombinant DNA Advisory Committee (RAC). Moreover, the USA Further Consolidated Appropriations Act of 2020 includes the Dickey-Wicker Amendment, which restricts federal funding of human embryos, along with another amendment that forbids the FDA from considering applications involving human germline genomic editing in clinical trials. Based on this regulatory framework, it is perhaps evident that at the time of writing this, no federal law exists in the USA which prohibits explicitly human cloning despite the federal laws and regulations which address funding and other indirect issues connected to cloning that exist (The Witherspoon Council on Ethics and the Integrity of Science, "Part Four: Cloning Policy in the United States", 2015b).

Due to the lack of federal regulation, certain states in the USA began imposing their own sets of laws. From 1997 through 2001, five states, including California, Louisiana, Michigan, Rhode Island, and Virginia, issued complete bans on human reproductive cloning and/or therapeutic cloning. In addition, Virginia and Michigan banned human genomic germline editing for any purpose. These regulations were permanent, while the other three states only imposed temporary (5–8 years) regulations and allowed the technology for research purposes (Sonfield, 2002). Since these regulations were set forth, most of the remaining states within the USA have provided their laws on human cloning. As of 2015, seven states prohibited cloning from producing children and human genomic germline modifications for biomedical research, while ten states prohibited cloning from producing children but allowed the technology for research purposes and legally required any cloned human embryos to be frozen or destroyed (The Witherspoon Council on Ethics and the Integrity of Science, "Appendix: State Laws on Human Cloning," 2015a). One state (Minnesota) appears to have regulatory guidelines that prohibit human genomic germline modifications for biomedical research, but does not address issues related to producing children. Lastly, several states have yet to provide the regulatory framework for the technology in any capacity, including Alabama, Alaska, Colorado, Florida, Nevada, and more.

Together with the National Academy of Medicine and the British Royal Society, the National Academy of Sciences has recently published a joint report on CRISPR and any heritable human genome editing biotechnologies in September 2020 (Adashi & Cohen, 2020). The report clearly emphasizes what is commonly seen across the globe: that CRISPR and other human germline editing methods should not be used for any clinical purpose in the modern day. However, the joint report also provides decision trees to aid in determining types of investigations, human trials, and settings that may allow these technologies. Furthermore, many renowned investigators and geneticists, including Dr. Jennifer Doudna and Dr. Feng Zheng, a neuroscientist at MIT, have praised the report, reiterating the fact that the CRISPR technology is still in its infancy, and not enough is known about its potential in human embryos (Angrist et al., 2020).

8.5 National Regulation in the UK

The majority of the legislature is consistent between the UK and most of the USA regarding CRISPR and human genomic germline technologies. This is primarily due to the updated guidelines of The Human Fertilisation and Embryology (HFE) Act of 1990 and the Regulations of 2008. The HFE Act of 1990 dictated that the research or clinical utilization of any gametes or embryos for any purpose is forbidden unless a license from the Human Fertilisation and Embryology Authority is obtained. Under specific circumstances (as dictated by Section 3ZA(4b) of the HFE Act), a license is always prohibited, including placing an embryo or gamete that has undergone nuclear or mitochondrial DNA modifications into a woman. Significantly, in 2008 and 2015, these regulations were updated, which ultimately banned heritable human genome editing but permitted "mitochondrial donations," or permission to conduct research involving mitochondrial manipulation of an embryo, to prevent mitochondrial disorders. This amendment has since become known as the "Mitochondrial Donation Regulations of 2015," and was particularly significant considering it made the UK the first country to allow HNGT for research purposes and challenged the global policy on human genomic germline editing entirely (The Human Fertilisation and Embryology (Mitochondrial Donation) Regulations, 2015).

8.6 National Regulation in China

Although the first attempt at human genomic germline modification was carried out in China and the first genetically modified babies were born there, Chinese stipulations on CRISPR and other genomic editing-related biotechnologies are relatively similar to the USA and the UK. The National People's Congress (NPC) and its Standing Committee have recently updated the Chinese human genomic and embryonic guidelines following the He Jiankui affair. In 2021, the NPC and Standing Committee adopted new laws on biomedical research known as the Civil Code (2021), the Basic Healthcare law, the Criminal Code (currently a proposed amendment), and the Biosecurity Law (a draft at the time of this writing) (Song & Joly, 2021). The Chinese Civil Code protects humans' rights and, interestingly, contains a separate section called *Book Four: Personality Rights* (人格权编, Ren Ge Quan Bian), which highlights:

> the rights to life, bodily rights, the right to health, the right of name, reputation, portrait, privacy, honour, and other personality interests… (Article 990 of the Chinese Civil Code, 2021; Song & Joly, 2021)

This is implemented to ensure the dignity of the people. In 2021, the drafting committee responsible for the content of the Civil Code added two articles; Article 1008, which required ethical approval for any clinical trials along with written consent, which required disclosure of the purpose, objectives, and potential risks of the trial, and Article 1009, which stated "any medical research activity associated with human gene and human embryo must comply with the relevant laws, administrative regulations and national regulation, must not harm individuals and violate ethical morality and public interest."

These articles were drafted quickly following He's scandalous investigation (Article 1008 of the Chinese Civil Code, 2021; Article 1009 of the Chinese Civil Code, 2021) and were met with both adulation and criticism by Chinese citizens and members of the Civil Code Society. Associate Director of the Civil Code Research Society of the Chinese Law Society acclaimed Article 1009, reiterating the necessity for clear and defined protection of personal liberty and personal dignity (Wang, 2020). However, other members of the drafting committee, such as the well-known civil law scholar Sun Xianzhong criticized these articles of the Civil Code, arguing that "civil rights start with birth and end with death" (Xianzhong, 2020). Numerous other members of the Society argued for or against the inclusion of Articles 1008 and 1009.

Despite the ongoing disagreements triggered by the amendments of the Chinese Civil Code, Articles 1008 and 1009 showed that the Chinese government found it necessary to regulate further embryonic research and genomic editing after the incident with researcher Dr. He Jiankui. Previously, the Chinese Civil Code was not particularly responsible for providing a legal basis for civil or criminal litigation in response to violations of the rules and laws on human genomic editing. This was deemed more responsible by the Supreme Court. However, the enactment of Articles 1008 and 1009 do provide such civil litigation for these crimes, as they now designate the illegal practice of human gene editing, cloning of human embryos, and "severe endangering of the security of human genetic resources" to be new types of crimes under the Chinese Criminal Code (Interpretation of the 11th Amendment to the Criminal Code of the P.R.C.: Seven Points to be Considered, 七大看点解析刑法修正案(十一)草案, 2020; Song & Joly, 2021).

8.7 National Regulation in France

Other than China, France is the only country that considers CRISPR and human genomic editing and embryonic research issues and regulation in its Civil and Criminal Codes. The Civil Code of France overtly bans any eugenic practices and modifications of human genetic material to alter offspring (Article 16-4 of the French Civil Code, 2004). The French Criminal Code also labels eugenics and reproductive cloning as crimes against humanity. Those in violation of such crimes can be imprisoned for up to 30 years, along with €7,500,000 fine (Article 214-2 of the French Penal Code, 2004).

8.8 International Legislation to Human Genomic Germline Modifications

Following Dr. Jiankui's research, the development of CRISPR, and the advent of CRISPR as a gene-editing tool to change the DNA of human embryos in 2015, international debate and concern regarding the use of biotechnologies to alter embryonic genetic material had resurfaced as a hot subject in terms of ethics, human rights, and the regulatory framework involved. As Britta C. van Beers alluded to in her recent article regarding international regulation of human genomic modifications, the fundamental aspect of existing international human rights on biomedical technologies is the idea that allowing genetic alterations in embryos or gametes would result in substantial changes to humankind that cannot be reversed (van Beers, 2020; Yamin, 2019). These concerns date back to as early as 1982 when the Parliamentary Assembly of the Council of Europe, which contains a European Convention on Human Rights, including 47 states, already considered in their "Recommendations on genetic engineering" Articles 2 and 3 on human rights, which detailed that "rights to life and to human dignity" imply a right to inherit genetic material which has not been modified (Recommendation on Genetic Engineering, 1982). Significantly, of these 47 states, only 29 have signed or ratified the convention for a variety of reasons, and other countries, such as the UK, declined ratification on the basis that the Convention was too restrictive (van Beers, 2020).

The Recommendations provided by the European Convention on Human Rights displayed the realization that a regulatory framework on altering human genomic germlines was necessary at the time. However, the early recommendations failed to deliver specific guidelines, resulting in questions on how strict laws and regulations on human genomic germline modifications should be. As such, a Committee of Ministers involved with the Parliamentary Assembly was prescribed for drafting such a framework under the same rights outlined in the European Convention on Human Rights (Parliamentary Assembly of the Council of Europe, 2017a; Recommendation on Genetic Engineering, 1982).

As discussed in section B on national regulation, the Oviedo Convention in 1997 was a pivotal event concerning the international agreement, in Europe, on the rules and guidelines of human genomic germline editing. Notably, the Oviedo Convention was the first international, legally binding policy in biomedical law. European countries agreed (as outlined in Article 13) that

> ...an intervention seeking to modify the human genome may only be undertaken for preventive, diagnostic or therapeutic purposes and only if its aim is not to introduce any modification in the genome of any descendants.

Article 13 of the Oviedo Convention undoubtedly stated a definite ban on human genomic germline editing for reproductive purposes (van Beers, 2020). This ban has also been endorsed by the Council of Europe's Parliamentary Assembly and the Committee on Bioethics (DH-BIO, 2015; Parliamentary Assembly of the Council of Europe, 2017b). This regulatory framework set forth between the 1980s and 1990s remains consistent today. The EU continues to associate human genomic germline alterations as a direct conflict with human rights and fundamental values. This legal framework also extends to patents on genomic germline alteration, which can be observed in Article 6 of the Biotech Directive and Article 16, which states that:

> ...patent law must be applied to respect the fundamental principles of safeguarding the dignity and integrity of the person.

Therefore, patents that may in any way affect the respect for human dignity are disallowed by the EU legislature (Case C-34/10, Oliver v. Greenpeace e.V, 2011).

As of the early 2000s, the international framework in the EU has further reiterated the relationship between human rights, dignity, and biotechnologies. This restatement is delineated first by the EU Charter of Fundamental Rights of 2000, which proclaims in its first chapter, *Human dignity*, the prohibition of eugenic practices for those aiming at selecting persons (EU Charter of Fundamental Rights, 2000). Furthermore, in 2001, the EU introduced human germline editing regulations specifically related to clinical trials in the "Clinical Trials Directive (2001)," followed by the "Clinical Trials Regulation (Regulation EU No 536/2014 of the European Parliament and of the Council of Apr. 16, 2014, on clinical trials on medical products for human use, and repealing, 2014)." These regulations clearly outlined that any clinical trials involving gene therapy that results in alterations to an individual's germline identity are prohibited, which effectively have blocked any thoughts of a safe introduction of these biotechnologies into the clinic (Regulation

EU No 536/2014 of the European Parliament and of the Council of April 16, 2014, on clinical trials on medical products for human use, and repealing, 2014).

Outside of European international law on human genomic germline editing, the United Nations Educational, Scientific and Cultural Organization (UNESCO), formed in 1945, declared in its "Universal Declaration on the Human Genome and Human Rights" that germline modifications are intertwined with human rights. UNESCO is a group with worldwide membership, including 193 different nations and 11 associate nations; however, the USA and Israel have not been members since 2017. In the UNESCO declaration, Article 1 outlined the organization's stance on human germline editing, stating that:

> The human genome underlies the fundamental unity of all members of the human family, as well as the recognition of their inherent dignity and diversity. In a symbolic sense, it is the heritage of humanity.

Article 24 of the declaration elaborates on UNESCO's stance on germline editing in humans as it provides advice on the "identification of practices that could be contrary to human dignities, such as germline intervention." These statements provide evidence that UNESCO defined germline interventions as biotechnology which contradicts human dignity. Additionally, a 2015 report was generated in response to new biotechnological developments such as CRISPR from the International Bioethics Committee (IBC). In the report, the IBC expressed its agreement and support of the regulatory framework provided by the Oviedo Convention's ban over genetic testing in personalized medicine, along with the use of human genomic germline editing. The IBC also added Article 1 of UNESCO's declaration, asserting that genetic changes made and transferred to offspring should be proscribed (Amelan, 2015).

8.9 CRISPR in Mental Disorders, Current WHO Guidelines, and Concluding Remarks

Another area of discussion with CRISPR and human genomic germline editing as a whole is the preemptive issues of which pathologies or conditions these biotechnologies can, or should, be applied to. As deliberated in the aforementioned text, a multitude of ethical issues exist regardless of the pathology when considering CRISPR as a potential tool. Treatment of heritable diseases such as sickle cell disease, hemophilia, muscular dystrophy, and cystic fibrosis, along with recent and ongoing clinical trials using CRISPR for the potential treatment of cancer and immune cell disorders, are just a few of the applications proposed (Baylis & McLeod, 2017; Kofler & Kraschel, 2018). However, another consideration is using CRISPR or other gene-editing biotechnologies to treat mental disorders and psychiatric illnesses. In this situation, the use of gene editing may pose additional layers of legal and ethical challenges, as it may involve modifications to our brain and overall nervous system, a system many humans believe defines us as an individual.

8.10 WHO Guidelines on CRISPR Utilization

To elaborate on the issues of potentially using CRISPR and other gene-editing technologies in humans, the recently published WHO guidelines (WHO, 2021) have updated their recommended policies for CRISPR utilization in research and clinical trials (WHO, 2021). The WHO recommendations stipulate that the world needs to more strictly regulate methods of human gene editing and aim to deter rogue-like science experiments involving alterations in the human genome. The updated WHO guidelines also discuss that usage and investigations of such biotechnologies should apply to the broader public and those in developing countries, and not just limited to the wealthy. For example, the WHO proposed a scenario in which researchers conduct a clinical trial on sickle cell gene editing in a sub-Saharan African population where the disease is highly prevalent. Even if the trial succeeded, the gene-editing treatment would be too expensive for most citizens of the country. Moreover, the WHO has started to document all studies involving genetic manipulation in their "Humane genome editing registry (WHO, 2021)," which was an idea launched in 2019 to make all information on clinical trials involving human genome editing accessible to the public. Finally, the WHO also stressed the need for increased international collaboration and attention to CRISPR and human genome editing, as each country must develop guidelines to ensure that ethical research is carried out with appropriate oversight, in hopes that this would further prevent unethical, illegal experiments such as Dr. Jiankui's in 2018.

Altogether, the current framework from the WHO, along with the National Academy of Science, National Academy of Medicine, and the British Royal Society, remain relatively consistent in their recommendations and guidelines regarding CRISPR and human genomic germline editing. These organizations agree that CRISPR is far too new, should not be used for clinical treatment at this time, and that whether or not it should be used in clinical trials is on a case-by-case basis. These organizations continue to stress the need for increased attention and international collaboration on these guidelines. Despite these efforts, many of these recommendations are merely suggestions. And these suggestions may not be enforceable without some form of international treaty or agreement with clear lines on what is or is not allowed and binds countries to their actions, a contract that entails the involvement of international organizations such as the United Nations to consolidate and, hopefully, unify such guidelines.

References

A declaration of the Académie nationale de médecine and of the Académie des sciences concerning the announcement made by Dr Hé Jiangkui (2018).

Adashi, E. Y., & Cohen, I. G. (2020). Heritable human genome editing: The international commission report. *JAMA, 324*(19), 1941–1942.

Amelan, R. (2015). *UNESCO panel of experts calls for ban on 'editing' of human DNA to avoid unethical tampering with hereditary traits*. http://www.unesco.org/new/en/social-and-human-sciences/themes/bioethics/sv0/news/unesco_panel_of_experts_calls_for_ban_on_editing_of_hu/

Angrist, M., Barrangou, R., Baylis, F., Brokowski, C., Burgio, G., Caplan, A., Chapman, C. R., Church, G. M., Cook-Deegan, R., Cwik, B., Doudna, J. A., Evans, J. H., Greely, H. T., Hercher, L., Hurlbut, J. B., Hynes, R. O., Ishii, T., Kiani, S., Lee, L. H., … Davies, K. (2020). Reactions to the National Academies/Royal Society report on heritable human genome editing. *The CRISPR Journal, 3*(5), 332–349. https://doi.org/10.1089/crispr.2020.29106.man

Araki, M., & Ishii, T. (2014). International regulatory landscape and integration of corrective genome editing into in vitro fertilization. *Reproductive Biology and Endocrinology, 12*(1), 108. https://doi.org/10.1186/1477-7827-12-108

Article 1008 of the Chinese Civil Code. (2021).

Article 1009 of the Chinese Civil Code. (2021).

Article 16-4 of the French Civil Code,'...Without prejudice to researches aiming at preventing and treating genetic diseases, there may be no alteration of the genetic characters with a view to changing the descent of a person'. (2004).

Article 214-2 of the French Penal Code, 'Carrying out any procedure designed to cause the birth of a child genetically identical to another person whether living or deceased is punished by thirty years' criminal imprisonment and a fine of €7,500,000.' (2004).

Article 990 of the Chinese Civil Code. (2021).

Baylis, F., & McLeod, M. (2017). First-in-human phase 1 CRISPR gene editing cancer trials: Are we ready? *Current Gene Therapy, 17*(4), 309–319. https://doi.org/10.2174/1566523217666171121165935

Case C-34/10, Oliver v. Greenpeace e.V. (2011).

Chinese Academy of Medical Sciences: 'Gene-edited Babies: Chinese Academy of Medical Sciences' Response and Action'. (2018, December). https://www.thelancet.com/action/showPdf?pii=S0140-6736%2818%2933080-0

Council, O. E. (1997). *Convention for the protection of human rights and dignity of the human being with regard to the application of biology and medicine: Convention on human rights and biomedicine*. COE.

Cyranoski, D., & Ledford, H. (2018). Genome-edited baby claim provokes international outcry. *Nature, 563*(7733), 607–608. https://doi.org/10.1038/d41586-018-07545-0

(DH-BIO), C. o. B. (2015). *Statement on Genome editing technologies*. https://rm.coe.int/168049034a

Doudna, J. (2019, April 29). He Jiankui. *Time Magazine*.

EU Charter of Fundamental Rights. (2000).

Greely, H. T. (2019). CRISPR'd babies: Human germline genome editing in the 'He Jiankui affair'. *Journal of Law and the Biosciences, 6*(1), 111–183. https://doi.org/10.1093/jlb/lsz010

Hendriks, A. (1997). Council of Europe-Convention for the protection of human rights and dignity of the human being with regard to the application of biology and medicine: Convention on human rights and biomedicine. *European Journal of Public Health, 4*, 89.

Interpretation of the 11th Amendment to the Criminal Code of the P.R.C.: Seven Points to be Considered, 七大看点解析刑法修正案(十一)草案. (2020, June 28). *Xinhua News*. https://www.ccdi.gov.cn/yaowen/202006/t20200628_220925.html

Isasi, R., Kleiderman, E., & Knoppers, B. M. (2016). Genetic Technology Regulation. Editing policy to fit the genome? *Science, 351*(6271), 337–339. https://doi.org/10.1126/science.aad6778

Kofler, N., & Kraschel, K. L. (2018). Treatment of heritable diseases using CRISPR: Hopes, fears, and reality. *Seminars in Perinatology, 42*(8), 515–521. https://doi.org/10.1053/j.semperi.2018.09.012

Lab, T. H. (2018). *About Lulu and Nana: Twin girls born healthy after gene surgery as single-cell embryos*. Retrieved July 9, 2021 from www.youtube.com/watch?v=th0vnOmFltc

Liu, S. (2020). Legal reflections on the case of genome-edited babies. *Global Health Research and Policy, 5*(1), 24. https://doi.org/10.1186/s41256-020-00153-4

Nuffield Council. *Supra note 38*.

Organization, W. H. (2019, July 26). *Statement on Governance and oversight of human genome editing*. Retrieved 7 July 2021 from https://www.who.int/news/item/26-07-2019-statement-on-governance-and-oversight-of-human-genome-editing

Parliamentary Assembly of the Council of Europe, Recommendation on Genetic Engineering, supra. (2017a).

Parliamentary Assembly of the Council of Europe, The use of New Genetic Technologies in Human Beings, § sub 3. (2017b).

Reardon, S. (2017). Genetic details of controversial 'three-parent baby' revealed. *Nature, 544*(7648), 17–18. https://doi.org/10.1038/nature.2017.21761

Recommendation on Genetic Engineering. (1982).

Regulation (EU) No 536/2014 of the European Parliament and of the Council of Apr. 16, 2014 on clinical trials on medical products for human use, and repealing. (2014).

Sonfield, A. (2002, February 1). Not waiting for congress to act, some states move to ban human cloning. *Guttmacher Institute, 5*(1). https://www.guttmacher.org/gpr/2002/02/not-waiting-congress-act-some-states-move-ban-human-cloning

Song, L., & Joly, Y. (2021). After He Jianku: China's biotechnology regulation reforms. *Medical Law International, 21*(2), 174–192. https://doi.org/10.1177/0968533221993504

Statement by the Organizing Committee of the Second International Summit on Human Genome Editing. (November 2018).

The Human Fertilisation and Embryology (Mitochondrial Donation) Regulations. (2015).

The Witherspoon council on Ethics and the Integrity of Science, "Appendix: State Laws on Human Cloning". (2015a). *The New Atlantis, Number 46*, 96–106. https://www.thenewatlantis.com/publications/appendix-state-laws-on-human-cloning

The Witherspoon Council on Ethics and the Integrity of Science, "Part Four: Cloning Policy in the United States". (2015b). *The New Atlantis, Number 46*, 74–90. https://www.thenewatlantis.com/publications/part-four-cloning-policy-in-the-united-states

van Beers, B. C. (2020). Rewriting the human genome, rewriting human rights law? Human rights, human dignity, and human germline modification in the CRISPR era. *Journal of Law and the Biosciences, 7*(1). https://doi.org/10.1093/jlb/lsaa006

Wang, Y. (2020, May 24). 'The 1260 Articles of the Civil Code: from Cradle to Cemetary; from value consensus to laws' 民法典1260条中的生老病死:从价值共识到法律规则. *Beijing News*. http://www.bjnews.com.cn/inside/2020/05/24/730856.html

WHO Expert Advisory Committee on Developing Global Standards for Governance and Oversight of Human Genome Editing. (2021). (Vol. License: CC BY-NC-SA 3.0 IGO.).

World's First Gene-Edited Babies Created in China. (2018). *The Guardian*. https://www.theguardian.com/science/2018/nov/26/worlds-first-gene-edited-babies-created-in-china-claims-scientist

Xianzhong, S. (2020). Regarding the Book of Personality Rights of the Civil Code submitted to the Twelfth Session of the Thirteenth Standing Committee of the National People's Congress. 孙宪忠:关于提交十三届全国人大常委会第十二次会议审议的民法典人格权编的审议意见. http://iolaw.cssn.cn/zt/sunxianzhong/

Yamin, R. A. A. E. (2019). *The right to design babies? Human Rights and bioethics*. https://www.openglobalrights.org/the-right-to-design-babies-human-rights-and-bioethics/

Yotova, R. (2020). Regulating genome editing under international human rights law. *International & Comparative Law Quarterly, 69*(3), 653–684. https://doi.org/10.1017/S0020589320000184

Chapter 9
Policy Recommendations Concerning CRISPR Germline Editing in Humans

9.1 Introduction

The conundrum of developing international treaties that respect the need for scientific progress while setting guardrails for scientific research requires the development of a framework that goes beyond current calls for recommendations. Creating and enforcing international treaties around CRISPR is easier said than done as the scientific work that encapsulates gene editing in general, and CRISPR, in particular, is increasingly international and collaborative. This makes developing and applying agreements involving human rights law extremely challenging. One reason for the challenging nature of this work is the disproportionate effect cultural differences play in ideas like personhood:

> The very concept of informed consent is almost meaningless in societies that stress the overriding importance of an individual's relationship with family and community. (Thomasma, 2001)

And indeed, the concept of bioethics itself:

> …bioethical principles are not uniform across the world and are strongly influenced by culture. (Schweikart, 2021)

This makes for a challenging environment concerning developing paths forward for a policy that includes the most comprehensive number of stakeholders. There is an additional caveat concerning policy considerations that flow from a discussion of bioethics, and that is that principles of bioethics do not typically provide a framework for legal discussions and policies:

> …while courts and policy-makers may be taking up bioethical issues in the context of human rights law, they will likely still use legal reasoning to make decisions, not the academic modes of reasoning traditionally used in the field of bioethics. This could be considered a limitation on effectuating or merging bioethical principles into human rights law. (Schweikart, 2021)

Thus, combining an anticipatory ethics that considers the widest number of stakeholders with an anticipatory governance framework will set the best stage for

M. W. Nestor, R. L. Wilson, *Anticipatory Ethics and The Use of CRISPR in Humans*, https://doi.org/10.1007/978-3-030-98368-0_9

successful policy concerning future uses of CRISPR-as we have outlined in earlier chapters of this manuscript. The purpose of using an anticipatory governance approach is to include stakeholder perspectives that encompass broad swaths of individuals and institutions, including stakeholders from the legal, scientific, and bioethics communities, amongst others:

> Anticipatory governance comprises the ability of a variety of lay and expert stakeholders, both individually and through an array of feedback mechanisms, to collectively imagine, critique, and thereby shape the issues presented by emerging technologies before they become reified in particular ways. Anticipatory governance evokes a distributed capacity for learning and interaction stimulated into present action by reflection on imagined present and future sociotechnical outcomes. (Barben et al., 2008 in Hackett, 2008)

An anticipatory governance approach based on a sound ethical and policy framework is also necessary because scientific discovery is unpredictable and often results far from the original research's stated intention. The stochastic and unpredictable nature of scientific discovery has led to many supporting the position that it is fruitless to set boundaries to shape the course for basic research. This was epitomized by Michael Polanyi as summarized by Bedessem and Ruphy (2019):

> Polanyi gives a somewhat lyrical form of this defense in his classical essay 'The Republic of Science' (1962, p. 62): 'Any attempt at guiding research towards a purpose other than its own is an attempt to deflect it from the advancement of science'. He adds that 'you can kill or mutilate the advance of science, you cannot shape it. For it can advance only by essentially unpredictable steps, pursuing problems of its own, and the practical benefits of these advances will be incidental and hence doubly unpredictable'. In Polanyi's view, the supposedly unpredictable nature of scientific development militates for an internal definition of research priorities...

In the case of setting policy with respect to the use of CRISPR, the prevailing notion in the scientific literature is that scientists should be largely self-governing but regulated by international guidelines "devised and monitored by scientists" (Akabayashi et al., 2018). This approach is a logical continuation of the Polanyi premise and tries to put the reins of governance, including anticipatory governance, in the hands of the scientists, in an attempt to keep from stifling scientific progress. However, we view this as an approach that may be too permissive concerning the lived experiences of a wide diversity of stakeholders. In the previous chapters of this manuscript, we have argued that the diverse perspectives of intersubjective life-worlds of current and future stakeholders are the most important antecedents to an appropriate and fitting ethical analysis of CRISPR. This also holds for the development of policy that shapes the use and application of this technology. Simply developing guidelines that are mostly "devised and monitored by scientists" are, as we have argued in Chap. 5, cutting out essential stakeholders and handing over policy to mostly unelected scientific elites. Bedessem and Ruphy (2019) argue that the occurrence of scientific innovation with new and unexpected discoveries flow best when the governance of scientific inquiry is taken out of the hands of scientists themselves and put into institutions that can provide oversight and guidance according to external considerations:

> In other words, in the absence of external incentives and even in a (ideal) realm of free individuals, scientists would favor 'exploitation' over 'exploration'. This suggests that some appropriate form of institutional control over scientific communities (e.g. forms that actively support risky science) might be preferable to 'blind delegation' in order to optimize the full flourishing of the unexpected. (Bedessem & Ruphy, 2019)

Indeed, the application of policies that bound CRISPR research need not inhibit its progress but may direct research funding and time investment to applications that meet the most pressing societal needs without tying up money and time in areas that, while interesting, may not be useful to the largest group of stakeholders. This is entirely in line with the linear model of innovation championed by Vannevar Bush (1945) that helped support the most expansive technological and research enterprise in modern history (Celeste et al., 2014). Guidelines and policy are not the enemies of progress. But what policies then are appropriate for shaping the future of CRISPR concerning the lived experience of future persons?

9.2 Policy Guidelines for Limiting CRISPR Use in Humans

The National Academies of Science (NAS) and the World Health Organization (WHO) have done tremendous and essential work developing guidelines for using CRISPR and CRISPR-related research. Indeed, a consortium of the most important figures in CRISPR-based research and development has based on the NAS guidelines, called for an out and out moratorium on germline genome editing (Lander et al., 2019). Given the scientific consensus, it seems that there will be a continued call for a moratorium at the third international summit in London in 2022 by the NAS. We support the WHO and NAS in their outreach and vision with respect to germline genome editing. However, the difficulty with guidelines and declarations is that these policy vehicles are non-binding and challenging to enforce.

If the scientific community has reached the consensus that it is unethical to perform CRISPR on human embryos, and the stakeholders in a diverse set of other groups agree, then the most impactful way to protect the lived experience of future persons is to create a binding set of international treaties. Given that there is consensus and the framework of guidelines from the WHO and NAS (as well as other sets of guidelines we outlined in Chap. 8), the United Nations Educational, Scientific and Cultural Organization (UNESCO) could be the appropriate body to debate and develop a set of oversight powers that would enable the shaping over time of binding regulations that could be turned into binding treaties:

> The international debate should include not only States but also scientists, NGOs, specialized international organizations, industry and persons carrying genetic diseases. This could be done under the auspices of UNESCO, the WHO or through an ad hoc conference. The conference could draft a binding treaty or a soft law declaration setting out internationally agreed standards applicable to genome editing and paving the way to a future binding instrument. Given the complex scientific character of the object of regulation, such a treaty could be drafted by an ILO [International Labor Organization]-style tripartite conference, including States, scientists and affected individuals. The conference could also draft a

> model law on genome editing to help guide the regulatory approaches of States with respect to the new technology given that domestic law and its effectiveness are the main safeguard for the lawful and ethical use of genome editing. Such a model law should set out a minimum threshold of protection for the embryo, acknowledging its human dignity and drawing inspiration from the key relevant human rights, such as the right to health, the right to benefit from science, the principle of no harm and/or the principle of the best interest of the future child. (Yotova, 2020).

While a direct recommendation on what such a set of treaties would look like or who might enforce such treaties is beyond the scope of this manuscript, it is essential to ask what principles policymakers should employ while developing binding treaties. Later in this chapter, we will introduce a tool that we believe could focus the potential for treaties while keeping scientific progress involving CRISPR-based research moving forward. By understanding the implications for enhancement, policymakers should (Fig. 9.1)

- Develop a set of international treaties that govern the use and application of CRISPR in humans.
- Engage large international workgroups that include a wide range of stakeholders, heavily skewed to non-scientists and non-experts.
- Define clearly the unintended consequences of CRISPR in humans and be open to an anticipatory analysis of these consequences even when told they are "far-fetched" by the scientific community.
- Create protocols for informed consent that consider international treaties concerning CRISPR.

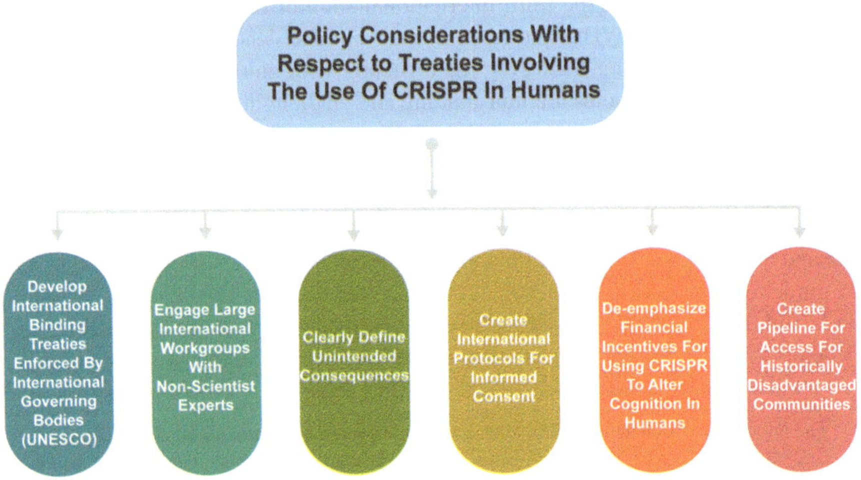

Fig. 9.1 Principles for policymakers to consider as they develop binding treaties concerning the use of CRISPR in humans. These principles involve including non-scientist experts and historically disadvantaged communities in the discussion around the development of treaties that protect human cognition from being altered by CRISPR approaches. These principles do not attempt to impede the progress of scientific research with respect to CRISPR or slow innovation

- Understand that CRISPR can, in specific contexts, represent a biological nuclear arms race and advocate to limit its application to basic research through policies that remove economic incentives for its use in particular contexts (e.g., altering cognition) in humans.
- Create a pipeline for access to CRISPR and its associated technologies that considers the challenges faced by historically disadvantaged groups and those from disadvantaged socioeconomic backgrounds.
- Be prepared to use international governing bodies (such as the United Nations) to enforce standards and sanction other countries for the inappropriate use of CRISPR in humans.

9.3 Artificial Intelligence as a Policy Analogue for CRISPR

Emerging technologies often follow similar trajectories concerning their ability to disrupt existing paradigms creatively and, as such, often can be examined and regulated using similar policy roadmaps (Groen & Walsh, 2013). CRISPR and artificial intelligence (AI) are similarly creatively disruptive emerging technologies that look to dominate scientific progress in the early part of the twenty-first century. AI is currently used in CRISPR-based research to improve the accuracy of gene editing (Yeager, 2019), design new therapeutic modalities (Louie et al., 2021), and even discover more anti-CRISPR proteins that act as "kill switches" (see Chap. 6) (Eitzinger et al., 2020; Gussow et al., 2020). Together with synthetic biology, CRISPR and AI have been suggested to form a "technology triad" that could be used with the evolution of quantum computing to engage in a radical new view of humanity through the lens of speciation (Bera, 2020). In the context of this "technology triad," AI and CRISPR share similarities in that they are both platform technologies that enable discoveries across disciplines (this could be said of synthetic biology to a degree as well). Here we adopt the idea that we can appropriate ideas about policy from one area of emerging technology that can be applied to another. Because of their similar trajectories and impact, we will adopt policy frameworks developed around AI and adapt them to CRISPR. The attempt to formulate policy must focus on an action related to what should be done and consider what outcomes we want to achieve. Through reflection, we can also reflect upon the intentions of these actions and how our actions and projected outcomes are related to the outcomes we are attempting to achieve.

According to Coeckelbergh (2020) (referring to policy for AI), several

> …questions also need to be answered in policy proposals: not only what should be done, when it should be done, how much should be done, by whom it should be done, and what the nature, extent, and urgency of the problem are. (Coeckelbergh, 2020)

In this case, the action we are interested in is developing policies for CRISPR. The outcome is the development of policy that considers the broadest possible group of stakeholders.

To move toward anticipatory governance and policy, we introduce and identify the "five rules for computing artifacts" and adopt them to CRISPR (Grodzinsky et al., 2015). The "five rules" have been developed by a community of scholars to attempt to address issues with computing artifacts (Grodzinsky et al., 2012; Miller, 2011). We employ the five rules as the first stage in developing an anticipatory policy. The five rules are introduced in the following way:

> As computing artifacts become increasingly complex, some have suggested that such artifacts greatly complicate issues of responsibility. In order to help deal with these complexities, we propose five rules as a normative guide for people who design, develop, deploy or use computing artifacts. Our aim is to reaffirm the importance of moral responsibility for these artifacts, and to encourage individuals and institutions to carefully examine their own responsibilities as they produce and use them. We do not claim that these rules are exhaustive; professionals, individuals and organizations may choose to take on more responsibility than we describe here. (Miller & Taddeo, 2017)

Two critical points are made in this passage. First, the five rules are important for emerging technologies other than AI because they encompass principles that stakeholders employ in their personal and professional capacities. Second, appropriation of the concepts from the five rules can be done flexibly and dynamically, giving the rules the framework of a living document. This allows their ready application to disruptive and innovative technologies such as CRISPR (Fig. 9.2).

- Rule 1: The people who design, develop, or deploy a computing artifact are morally responsible for that artifact and for the foreseeable effects of that artifact. This responsibility is shared with other people who design, develop, deploy, or knowingly use the artifact as part of a sociotechnical system.
  - *Modified Rule 1: The people and institutions who design, develop, or deploy CRISPR are morally responsible for that artifact and for the foreseeable effects of that artifact. This responsibility is shared with other people who design, develop, deploy, or knowingly use the artifact as part of a sociotechnical system.*
- Rule 2: The shared responsibility of computing artifacts is not a zero-sum game. The responsibility of an individual is not reduced simply because more people become involved in designing, developing, deploying, or using the artifact. Instead, a person's responsibility includes being answerable for the behaviors of the artifact and the artifact's effects after deployment, to the degree to which these effects are reasonably foreseeable by that person.
  - *Modified Rule 2: The shared responsibility of CRISPR artifacts is not a zero-sum game. An individual's responsibility is not reduced simply because more people become involved in designing, developing, deploying, or using the artifact. Instead, a person's or institution's responsibility includes being answerable for the behaviors of the artifact and the artifact's effects after deployment, to the degree to which these effects are reasonably foreseeable by that person or institution.*

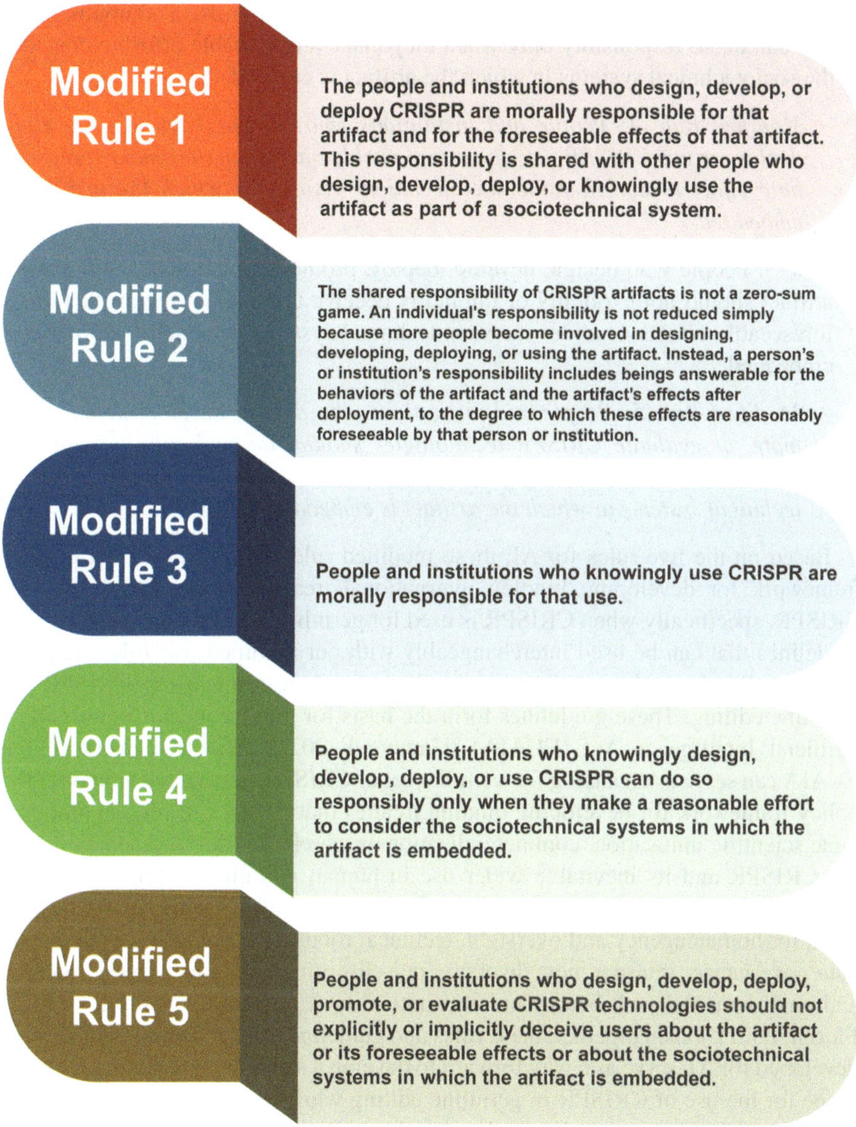

Fig. 9.2 Modification of the "five rules for computing artifacts" applied to CRISPR technology provides a starting set of guidelines for developing long-term policy with respect to the use of CRISPR in humans. The modified five rules focus attention on the responsibilities of stakeholders with respect to how CRISPR is used and evaluated

- Rule3: People and institutions who knowingly use a particular computing artifact are morally responsible for that use.
  - *Modified Rule 3: People and institutions who knowingly use CRISPR are morally responsible for that use.*

- Rule 4: People who knowingly design, develop, deploy, or use a computing artifact can do so responsibly only when they make a reasonable effort to consider the sociotechnical systems in which the artifact is embedded.
  - *Modified Rule 4: People and institutions who knowingly design, develop, deploy, or use CRISPR can do so responsibly only when they make a reasonable effort to consider the sociotechnical systems in which the artifact is embedded.*
- Rule 5: People who design, develop, deploy, promote, or evaluate a computing artifact should not explicitly or implicitly deceive users about the artifact or its foreseeable effects or about the sociotechnical systems in which the artifact is embedded.
  - *Modified Rule 5: People and institutions who design, develop, deploy, promote, or evaluate CRISPR technologies should not explicitly or implicitly deceive users about the artifact or its foreseeable effects or about the sociotechnical systems in which the artifact is embedded.*

Based on the five rules for AI, these modified rules should serve as a guiding framework for developing binding international treaties concerning the use of CRISPR, specifically when CRISPR is used for germline editing. Adding a layer of guidelines that can be used interchangeably with our modified five rules creates a robust toolkit for policy and treaty development concerning using CRISPR for germline editing. These guidelines form the basis for the European Commission's Artificial Intelligence Act (EUAIA) (Gaumond, 2021). The framework of the EUAIA can serve as an analog for a similar set of CRISPR Acts which would set the policy framework for developing binding treaties that would still respect and promote scientific innovation, commercialization, research, and development concerning CRISPR and its inevitable wider use in human germline editing. The seven guidelines, called the "Ethics Guidelines For Trustworthy AI" (Saif, 2020), include a call for human agency and oversight, technical robustness and safety, privacy and data governance, transparency, diversity, non-discrimination and fairness, societal and environmental well-being, and accountability (European Commission, 2021; Floridi, 2021). Taken together, these rules and guidelines, taken from the framework developed for AI, serve as a way forward to develop a scientific and regulatory landscape for the use of CRISPR in germline editing while keeping scientific progress and yet creating strong guardrails through the definition of binding "red lines" for research.

These frameworks can also be combined with ideas from international policy approaches like the Treaty on the Non-Proliferation of Nuclear Weapons and other biological, chemical, and nuclear treaties (Nuclear Threat Initiative, 2021). Given the view of some in the US government that gene editing is a "weapon of mass destruction and proliferation" (Regalado, 2016; Servick, 2017) and the potential use of CRISPR in biowarfare (Gronlund, 2021) and other types of "CRISPR wars" (Werner, 2019), it is essential to develop treaties and red lines early in the development of this technology to keep innovation concerning CRISPR and

CRISPR-mediated research and development progressing. In addition, these international treaties will ensure that CRISPR is held in high regard by the lay public so that CRISPR research maintains adequate political support, translating into ample funding of research from government agencies.

9.4 Additional Policy Analogs for Germline CRISPR Editing Governance

Nonetheless, the governance of human gene editing via CRISPR requires more profound policy guidelines that go beyond informed consent, short- and long-term risk, and equitable access to the technology. Indeed, as the applications of CRISPR increase, so will the complexity of the policy considerations. Juengst (2021) considers the concept of preventions as opposed to treatments (which most policy frameworks take into consideration) concerning both somatic and germline editing as an essential additional consideration for future policy frameworks:

> Each form of prevention sends us in a different direction for guidance: phenotypic prevention, to our emerging experience with preemptive genetic medicine; genotypic prevention, to our history of efforts to control gene flow; and preventive strengthening, to the translational pipelines of beneficial genomic variant research. What should integrate and ground these efforts is a renewed resolve to never again allow invidious genetic value judgments to undermine our commitment to our common human moral equality in the face of our biological diversity. (Juengst, 2021)

The need for complex policy that tackles both treatments and prevention as distinct aspects of germline CRISPR editing will clarify which aspects of CRISPR-based research need binding international agreements and which do not.

Scott Schweikart (2019) provides two more historical policy frameworks that may serve as valuable analogs for the anticipatory governance of CRISPR germline editing in humans. The first is the Asilomar conference of 1975. Asilomar was convened to consider the potential health and environmental risks associated with the use of recombinant DNA (rDNA) and resulted in a worldwide moratorium on rDNA research. The moratorium was eventually lifted due to the lack of risks associated with rDNA use. Still, it resulted in the National Institutes of Health Biosafety and Recombinant DNA Policy concerning recombinant research. Interestingly, Schweikart notes that the conference failed to consider the broadest groups of stakeholders possible and, as such, excluded to a great extent "…governance applications and broader societal and ethical aspects of the technology." (Schweikart, 2019). This underscores the pressing need for an anticipatory ethics and governance framework for policy considerations with respect to the use of CRISPR in germline applications for humans.

The second policy scheme that Schweikart highlights is from the German Ethics Council (GEC). The GEC convened in 2019 and released a set of guidelines on human germline genome editing. What is vital for current policymakers is the call from the GEC to develop a moratorium on human germline editing that is an internationally binding agreement. Like the guidelines put forth by the National Academy of Sciences and the Nuffield Council, the GEC included the expertise from large groups of stakeholders outside the scientific community and considered "ethical and societal considerations beyond a risk-benefit analysis" (Schweikart, 2019).

For this discussion, in light of our support for international binding treaties around certain aspects of the use of CRISPR in humans, the critical policy consideration that arose out of the GEC is the:

> ...creation of an international agency that would evaluate the 'scientific, medical, ethical, legal, societal and political implications of germline interventions in humans'. (Schweikart, 2019)

This international agency should be housed in the United Nations (UN) and work closely with member states and the World Health Organization (WHO) to develop binding agreements and policy structures that synthesize diverse sets of stakeholders' interests and concerns. A limitation of the GEC is that although it did use the perspectives of diverse expertise in developing its guidelines, the process and policy considerations were largely devoid of the perspective of the lay public, particularly the least represented and most vulnerable populations within the lay public. The UN and WHO have the ability to integrate the lived experiences of these types of stakeholders along with traditional experts, elites, and state actors/institutions concerning CRISPR germline editing. This will be important in the governance of moratoriums on CRISPR-related research in the future:

> If such a prudent path (ie, one that appreciates the technology's risks in recommending a moratorium but is also mindful of its benefits in making the moratorium nonsticky and modifiable) can become a true international consensus on how to approach genome editing (particularly germline editing), such a consensus can likely become, in the absence of a binding international agreement, the best possible way to mitigate the technology's risk on a global scale. (Schweikart, 2019)

Given the uncertainty around the use of CRISPR in human germline editing, the current calls for a moratorium are prudent but not sustainable. As was the case with the Asilomar conference of 1975, research on rDNA continued even with the de facto moratorium. There was significant political pushback, including from "dissident scientists" who disagreed with Asilomar (Fredrickson, 1990). Notably, the Asilomar moratorium did provide time for policymakers to develop frameworks for responsible rDNA research, but it did not stop this research. However, it is critical to point out that the pace and quantity of research data and publication output were a fraction of what they are today. This type of policymaking via moratoriums or de facto moratoriums buttressed by declarations and guidelines will be quickly outpaced in the current research environment.

9.5 The CRISPR Cognition Genes Database

Within the context of our argument throughout this manuscript, we do not think that a general moratorium on CRISPR germline editing in humans is appropriate or enforceable. However, we advocate for a moratorium on editing genes that are implicated in human cognition. The overarching theme of this manuscript lays out the case for why we think this is the case. Beyond the philosophical and other ethical arguments for this position, a moratorium for this type of research is more straightforward to enforce than a general moratorium on all germline CRISPR editing. Instead of the database of CRISPR experiments proposed by the WHO guidelines, governing bodies could present a dynamic and updating list of genes based on human and animal studies that give researchers a clear red line for CRISPR-based experiments in humans. The list could be validated, and genes could be added and subtracted by scientific peer review. We propose to call this *The CRISPR-cognition genes database (CCGD)* (Fig. 9.3). The list and database would also be subject to public peer review, and discussion and input on the list would be gathered from diverse stakeholders. The results of these reviews and input would be used to feedback into the downstream review process. This is not to prevent research from moving forward on work involving CCGD genes, but to put the proper policy and regulatory frameworks to support ongoing research while assuring diverse stakeholders. The process that would inform the CCGD encourages active and dynamic feedback from stakeholders from diverse perspectives and allows for transparency and multiple points of peer review.

This type of system is manageable from a scale perspective and amenable to multiple points of entry for stakeholder involvement as genes are added or removed from the database. CRISPR-based animal models and editing in human cells using a somatic approach based on this theoretical gene database would be permissible, thereby keeping open the door to progress in research on cognition, learning, and memory. According to their legal customs and culture, the CCGD would allow international partners to enforce a focused and bounded moratorium on germline CRISPR editing. The CCGD could also help shape the discussion around what areas of germline editing would be permissible in the right human rights legal and regulatory environment, opening the possibility for some types of germline editing and even enhancement (such as viral resistance or other potentially medically important modalities). Finally, the CCGD is an easily understood research tool in the context of databases that scientists commonly use for high-throughput screening, drug discovery, and CRISPR-screens, like iCSDB and GenomeCRISPR, among others (Choi et al., 2020; Rauscher et al., 2016). The familiarity will encourage active adoption and participation by both scientists and non-scientists alike.

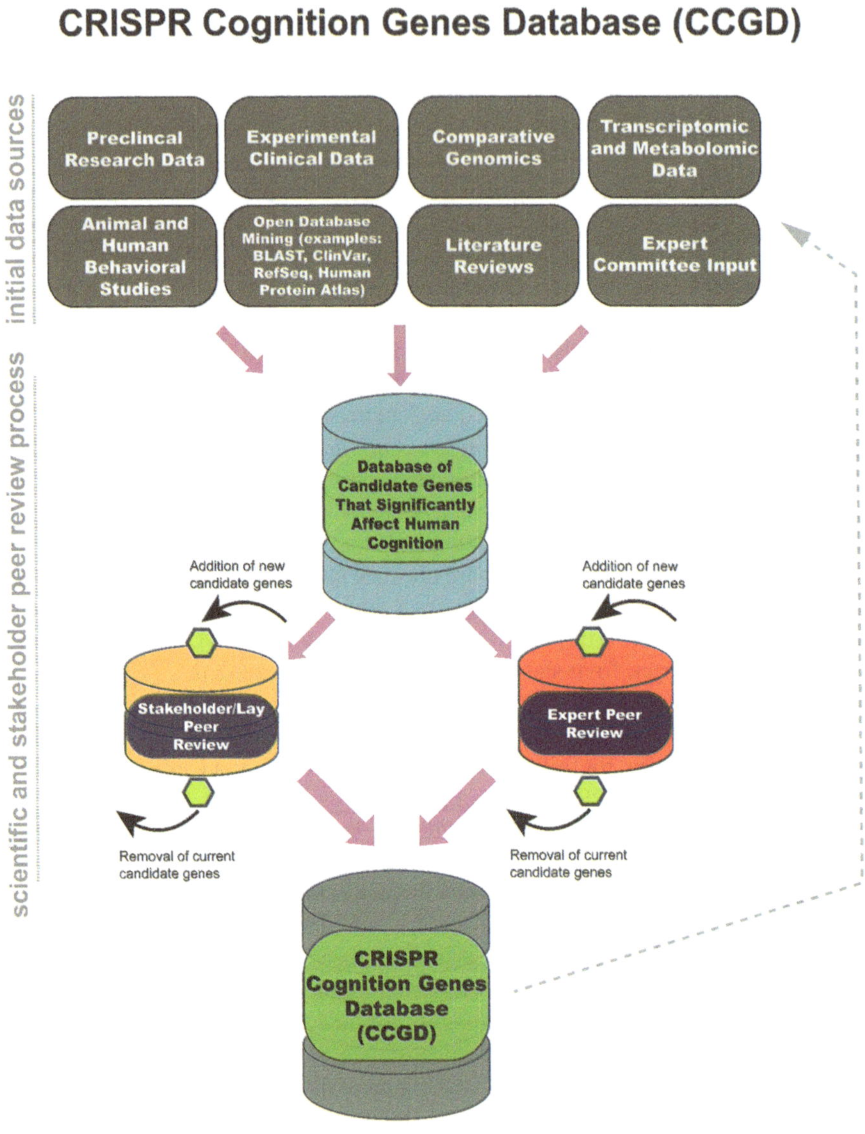

Fig. 9.3 The CRISPR Cognition Genes Database (CCGD) relies on the scientific and lay stakeholder community working together to aggregate data from diverse scientific sources to create a living database of genes determined to significantly affect human cognition. This database will undergo peer review, and genes can be added and removed as additional data is obtained. This recursive database will be constantly reviewed and will feedback into and inform additional clinical and preclinical data collection. The genes in this database should undergo further scrutiny and review by stakeholders, and policy should be set to potentially prevent CRISPR alterations of genes in the CCGD

References

Akabayashi, A., Nakazawa, E., & Caplan, A. L. (2018, December 12). *Gene editing: Who should decide?* Retrieved from https://www.nature.com/articles/d41586-018-07720-3

Bedessem, B., & Ruphy, S. (2019). Scientific autonomy and the unpredictability of scientific inquiry: The unexpected might not be where you would expect. *Studies in History and Philosophy of Science Part A, 73*, 1–7. https://doi.org/10.1016/j.shpsa.2018.08.002

Bera, R. K. (2020). Synthetic biology, artificial intelligence, and quantum computing. *Synthetic Biology–New Interdisciplinary Science*. https://doi.org/10.5772/intechopen.83434

Bush, V. (1945). *Science, the endless frontier: A report to the president*. United States Government Printing Office.

Celeste, R. F., Griswold, A., Straf, M. L., … Celeste, R. F. (2014). *Furthering Americas research enterprise*. National Academies Press.

Choi, A., Jang, I., Han, H., Kim, M., Choi, J., Lee, J., … Lee, S. (2020). ICSDB: An integrated database of CRISPR screens. *Nucleic Acids Research, 49*(D1). https://doi.org/10.1093/nar/gkaa989

Coeckelbergh, M. (2020). *AI ethics*. MIT Press.

Eitzinger, S., Asif, A., Watters, K. E., Iavarone, A. T., Knott, G. J., Doudna, J. A., & Minhas, F. U. (2020). Machine learning predicts new anti-CRISPR proteins. https://doi.org/10.1101/854950

European Commission. (2021). *Proposal for a Regulation of the European Parliament and of the Council Laying Down Harmonised Rules on Artificial Intelligence (Artificial Intelligence Act) and Amending Certain Union Legislative Acts COM/2021/206 final*. Retrieved from https://stip.oecd.org/stip/policy-initiatives/2019/data/policyInitiatives/24384

Floridi, L. (2021). The European Legislation on AI: A brief analysis of its ... Retrieved from https://link.springer.com/content/pdf/10.1007/s13347-021-00460-9.pdf

Fredrickson, D. S. (1990). *Asilomar: The end of the beginning*. Institute of Medicine.

Gaumond, E. (2021, June 9). *Artificial Intelligence Act: What is the European approach for AI?* Retrieved from https://www.lawfareblog.com/artificial-intelligence-act-what-european-approach-ai

Global enterprise to strengthen Non-Proliferation and disarmament. (2022, February 10). Retrieved April 20, 2022, from https://www.nti.org/about/programs-projects/project/global-enterprise-strengthen-non-proliferation-and-disarmament/

Grodzinsky, F. S., Miller, K., & Wolf, M. J. (2012). Moral responsibility for computing artifacts. *ACM SIGCAS Computers and Society, 42*(2), 15–25. https://doi.org/10.1145/2422509.2422511

Grodzinsky, F. S., Miller, K. W., & Wolf, M. J. (2015). Developing automated deceptions and the impact on trust. *Philosophy & Technology, 28*(1), 91–105. https://doi.org/10.1007/s13347-014-0158-7

Groen, A., & Walsh, S. (2013). Introduction to the field of emerging technology management. *Creativity and Innovation Management, 22*(1), 1–5. https://doi.org/10.1111/caim.12019

Gronlund, K. (2021, April 5). *Genome editing and the future of biowarfare: A conversation with Dr. Piers Millett*. Retrieved from https://futureoflife.org/2018/10/12/genome-editing-and-the-future-of-biowarfare-a-conversation-with-dr-piers-millett/

Gussow, A. B., Park, A. E., Borges, A. L., Shmakov, S. A., Makarova, K. S., Wolf, Y. I., … Koonin, E. V. (2020). Machine-learning approach expands the repertoire of anti-CRISPR protein families. *Nature Communications, 11*(1). https://doi.org/10.1038/s41467-020-17652-0

Hackett, E. J. (2008). *The handbook of science and technology studies*. MIT Press.

Juengst, E. T. (2021). "Prevention" and human gene editing governance. *AMA Journal of Ethics, 23*(1), E49–E54. https://doi.org/10.1001/amajethics.2021.49

Lander, E. S., Baylis, F., Zhang, F., Charpentier, E., Berg, P., Bourgain, C., … Winnacker, E. (2019, March 13). *Adopt a moratorium on heritable genome editing*. Retrieved from https://www.nature.com/articles/d41586-019-00726-5

Louie, W., Shen, M. W., Tahiry, Z., Zhang, S., Worstell, D., Cassa, C. A., … Gifford, D. K. (2021). Machine learning based CRISPR gRNA design for therapeutic exon skipping. *PLoS Computational Biology, 17*(1). https://doi.org/10.1371/journal.pcbi.1008605

Miller, K. W. (2011). Moral responsibility for computing artifacts: "The rules". *IT Professional, 13*(3), 57–59. https://doi.org/10.1109/mitp.2011.46

Miller, K., & Taddeo, M. (2017). *The ethics of information technologies*. Routledge.

Rauscher, B., Heigwer, F., Breinig, M., Winter, J., & Boutros, M. (2016). GenomeCRISPR - a database for high-throughput CRISPR/Cas9 screens. *Nucleic Acids Research, 45*(D1). https://doi.org/10.1093/nar/gkw997

Regalado, A. (2016). *Top U.S. intelligence official calls gene editing a WMD threat*. Retrieved from https://www.technologyreview.com/2016/02/09/71575/top-us-intelligence-official-calls-gene-editing-a-wmd-threat/

Saif, I. (2020, April 9). *'Trustworthy AI' is a framework to help manage unique risk*. Retrieved from https://www.technologyreview.com/2020/03/25/950291/trustworthy-ai-is-a-framework-to-help-manage-unique-risk/

Schweikart, S. J. (2019). What is prudent governance of human genome editing? *AMA Journal of Ethics, 21*(12), E1042–E1048. https://doi.org/10.1001/amajethics.2019.1042

Schweikart, S. J. (2021). Global regulation of germline genome editing: Ethical considerations and Application of International Human Rights Law, 43 Loy. L.A. Int'l & Comp. L. Rev. 279 (2021). Available at: https://digitalcommons.lmu.edu/ilr/vol43/iss3/6

Servick (2017, December 8). *CRISPR-a weapon of mass destruction?* Retrieved from https://www.sciencemag.org/news/2016/02/crispr-weapon-mass-destruction

Thomasma, D. C. (2001). Evolving bioethics and international human rights. In *Autonomy and human rights in health care international library of ethics, law, and the new medicine* (pp. 11–24). https://doi.org/10.1007/978-1-4020-5841-7_1

Werner, E. (2019). *The Coming CRISPR Wars: Or why genome editing can be more dangerous than nuclear weapons*. Retrieved from https://www.researchgate.net/publication/330507499_The_Coming_CRISPR_Wars_Or_why_genome_editing_can_be_more_dangerous_than_nuclear_weapons

Yeager, A. (2019). *Could AI make gene editing more accurate?* Retrieved from https://www.the-scientist.com/the-literature/could-ai-make-gene-editing-more-accurate-65781

Yotova, R. (2020). Regulating genome editing under international human rights law. *International & Comparative Law Quarterly, 69*(3), 653–684. https://doi.org/10.1017/s0020589320000184

Chapter 10
Conclusions

In this manuscript, we have attempted to develop a conversation between a philosopher and a scientist that examines some of the open questions around the use of CRISPR in humans from an anticipatory ethics perspective. We started this conversation from the perspective that the use of CRISPR in humans is a welcome addition to the wealth of biomedical approaches available to help ameliorate disease and reduce suffering. Although, for now, there is a de facto moratorium on the use of CRISPR for germline editing in humans, we believe that there will be reason to slowly and carefully remove any moratorium placed on CRISPR work in humans, akin to the path followed by the Asilomar conference of 1975 for recombinant DNA research. For this reason, we explored the possibility that complex mental traits could be edited using CRISPR technologies in the future. We acknowledge that this possibility is way beyond the capability of CRISPR in its current form and use. It may be a very long time before such a possibility is realized. But this is the central tenet of anticipatory ethics, to explore the technologically possible before it is realized. We used a small set of stakeholders to explore some of these ideas for our analysis. In the context of this potential future use, we tried to use concepts from anticipatory ethics, and a framework for our anticipatory ethical analysis was based on phenomenology, postphenomenology, and classical bioethics approaches.

We have argued that in the best interest of future persons' lived experiences, and it is important to involve the most comprehensive set of diverse stakeholder perspectives when developing policy related to germline CRISPR editing in humans. What emerges as a result of this analysis are several important insights. There is a need to think about the future when we start thinking about gene-editing technology. If we start from where we are today and reflect upon how to build a better future in the context of gene editing, it seems clear that a wide range of stakeholder views need to be taken into consideration. Each stakeholder can imagine a future that has a different configuration based on their life-worlds and lived experiences. Once a future has been projected and goals are set,

M. W. Nestor, R. L. Wilson, *Anticipatory Ethics and The Use of CRISPR in Humans*, https://doi.org/10.1007/978-3-030-98368-0_10

> developing systematic ways of thinking about [those] goals provide specific guidance on which directions to pursue and how to pursue them. (Montfort, 2017)

There is also a need to project different ethical narratives, including creatively projecting how ethical principles, codes of ethics, and anticipatory ethics frameworks can be applied to the future development of gene-editing technology and CRISPR. Finally, there is a need to move beyond declarations and guidelines and into binding international agreements to assure the best possible rollout of CRISPR and the most significant impact of this amazing technology for the greatest number of stakeholders. The future for CRISPR and biomedical innovation, in general, is bright. It is not surprising that this technology, among its peers in artificial intelligence, quantum computing, and synthetic biology, is ushering in a new, well-deserved era of technological optimism (Economist, 2021). But with this optimism, there needs to be a slow and methodical approach from the scientific and policy communities. CRISPR can contribute so much to our understanding of human disease and potential for curing disease, but as Paul Berg stated about the aftermath of Asilomar:

> …there is a lesson in Asilomar for all of science: the best way to respond to concerns created by emerging knowledge or early-stage technologies is for scientists from publicly-funded institutions to find common cause with the wider public about the best way to regulate — as early as possible. (Berg, 2008)

Common cause around a future we all have a stake in, and one in which we protect the lived experience of future persons is the path forward to developing policy programs that allow for innovation and scientific progress for CRISPR-based research while protecting the most vulnerable among us. There is room for all to have a seat at the table, and by providing seats for diverse sets of stakeholders, we ensure that we will all be as excited as researchers, scientists, clinicians, and patients are for a future in which CRISPR will pave the way for even more innovations in biomedicine and gene therapy. There will be a next CRISPR. Who knows what it will be called. We know that the progress of science, when coupled with the best interests of all of us, represents the best humanity has to offer for both current and future persons.

References

Berg, P. (2008). Asilomar 1975: DNA modification secured. *Nature, 455*(7211), 290–291. https://doi.org/10.1038/455290a

Economist. (2021). *Why a dawn of technological optimism is breaking*. Retrieved from https://www.economist.com/leaders/2021/01/16/why-a-dawn-of-technological-optimism-is-breaking

Montfort, N. (2017). *The future*. MIT Press.

Index

M. W. Nestor, R. L. Wilson, *Anticipatory Ethics and The Use of CRISPR in Humans*, https://doi.org/10.1007/978-3-030-98368-0

The manufacturer's authorised representative in the EU is Springer Nature Customer Service Centre GmbH, Europaplatz 3, 69115 Heidelberg, Germany. If you have any concerns regarding our products, please contact ProductSafety@springernature.com

Printed and bound by CPI Group (UK) Ltd, Croydon, CR0 4YY
30/09/2026
02225309-0016